The Importance of Vitamins to Human Health

The Importance of
Vitamins to Human Health

Proceedings of the IV Kellogg Nutrition Symposium
held at the Royal College of Obstetricians and Gynaecologists,
London, on 14–15 December, 1978

Edited by
T.G. Taylor
Rank Professor of Applied Nutrition,
School of Biochemical and Physiological Sciences,
University of Southampton

MTP PRESS LIMITED·LANCASTER·ENGLAND
International Medical Publishers

Published by
MTP Press Limited
Falcon House
Lancaster, England

British Library Cataloguing in Publication Data
Importance of Vitamins to Human Health.
(Conference), London, 1978
The importance of vitamins to human health.
1. Vitamins in human nutrition – Congresses
I. Taylor, Thomas Geoffrey, *b. 1918*
II. Kellogg Company of Great Britain
613.2'8   TX553.V5
ISBN 0-85200-268-8

Printed in Great Britain by Mather Bros (Printers) Ltd, Preston

Phototypesetting by Rainbow Graphics, Liverpool

Contents

List of contributors

G. C. ARNEIL
Department of Child Health,
University of Glasgow
Royal Hospital for Sick Children

A. E. BENDER
Department of Food Science and Nutrition,
Queen Elizabeth College,
University of London

G. B. BRUBACHER
Department of Vitamin and Nutritional
Research,
F. Hoffmann-La Roche & Co. Ltd.,
Basle, Switzerland

I. CHANARIN
Section of Haematology,
MRC Clinical Research Centre,
Harrow

J. W. T. DICKERSON
Division of Nutrition and Food Science,
Department of Biochemistry,
University of Surrey, Guildford

A. N. EXTON-SMITH
Department of Geriatric Medicine,
University College Hospital Medical School,
London

F. M. HASSAN
Department of Human Nutrition,
London School of Hygiene and
Tropical Medicine

A. V. HOFFBRAND
Department of Haematology,
The Royal Free Hospital, London

J. KELLEHER
University Department of Medicine,
St James's Hospital, Leeds

D. E. M. LAWSON
MRC Dunn Nutritional Laboratory,
University of Cambridge

M. S. LOSOWSKY
University Department of Medicine,
St James's Hospital, Leeds

G. A. J. PITT
Department of Biochemistry,
University of Liverpool

Hilary J. POWERS
Department of Human Nutrition,
London School of Hygiene and
Tropical Medicine

S. B. ROSALKI
Department of Chemical Pathology,
The Royal Free Hospital, London

C. J. SCHORAH
Department of Chemical Pathology,
University of Leeds

T. G. TAYLOR
School of Biochemical and
Physiological Sciences,
University of Southampton

D. I. THURNHAM
Department of Human Nutrition,
London School of Hygiene and
Tropical Medicine

R. A. WIGGINS
Department of Industry,
Laboratory of the Government Chemist,
London

Foreword

The Kellogg Company ranks among the world's leading food manufacturers, and it follows, therefore, that our corporate policies are important to human health. Indeed food manufacturers, as a combined industrial force, must bear a major responsibility for the health of mankind because commercially processed foods are increasingly an indispensable part of the mosaic of human nutrition.

This is particularly true in advanced industrialised societies. Here, in Great Britain, 40% of the people live in major conurbations and 41% of the food they eat is either pre-cooked or preserved, compounded or frozen, dehydrated or concentrated, or modified in some other way to satisfy a consumer need or preference. These preferences are communicated to the manufacturer through the competitive forces of the market, and are then translated into products in their most attractive and saleable form.

However, it is questionable how far consumer choice, depending largely on sight and taste, can be relied upon to ensure a correctly balanced and nutritionally adequate diet. The probable answer is that if we all relied, solely, on our senses and our appetites, many of us would be suffering from some form of nutritional imbalance.

A serious nutritional responsibility therefore rests with the modern food manufacturer. We, at Kellogg's, are conscious of the need, not only to make the foods we produce attractive to the purse and palate, but to ensure that they make a sound contribution in nutritional terms.

The dietary role of vitamins is therefore of pressing importance to us; and we are currently fortifying our products in line with the best available advice. But the progress of knowledge in this sphere is relatively slow moving, and we organised the symposium which is the subject of this book because we believed the time to be opportune for an assembly of leading nutritionists, medical practitioners, specialists and researchers in the field so that the latest advances could be explored and discussed.

We believe that the papers here presented will constitute an important contribution to the science of human nutrition.

In conclusion we wish to express our appreciation for the considerable assistance of various speakers and their associates given to us in the planning and organization of this Symposium. In particular, we wish to thank the two chairmen of the Symposium and the Editor of these Proceedings, respectively Professor I. MacDonald, Guy's Hospital, London; Dr M.R. Turner, British Nutrition Foundation, London and Professor T.G. Taylor, Southampton.

G.D. ROBINSON
CHAIRMAN AND MANAGING DIRECTOR
Kellogg Company of Great Britain Ltd.

Preface

Doctors and professional nutritionists have always been aware of the importance of vitamins to health but the components of the diet that have been most exposed to the glare of publicity in recent years have tended to be substances other than vitamins, in particular, fat, fibre, sugar and food additives. The publication of this symposium, therefore, is particularly welcome at the present time in restoring the balance.

Vitamin deficiency conditions are traditionally associated with poor, restricted diets and although the classical vitamin-deficiency diseases of pellagra, scurvy and beri-beri are mercifully rare in Britain, the same circumstances that combined together to produce these diseases in the past — poverty and ignorance — still operate to a lesser degree to-day in the vulnerable groups of the community, particularly in the elderly.

Vitamins of the B-group tend to have a similar distribution in foods, so that when deficiencies occur, several members of the group are likely to be implicated, and conditions of multiple deficiency result. These deficiency conditions are seldom severe enough to induce the deficiency symptoms typical of the individual vitamins. The manner in which marginal, multiple vitamin deficiency conditions express themselves are fairly non-specific and may be no more than a general malaise. These problems and the factors that contribute to them are discussed by Dr Exton-Smith, Dr Schorah, Dr Kelleher and by Dr Brubacher.

How then are these sub-clinical deficiency conditions to be diagnosed? Fortunately there are biochemical tests by which these marginal vitamin deficiencies can be identified and these tests are discussed by Dr Brubacher, Dr Thurnham, Dr Rosalki, Dr Kelleher and Professor Hoffbrand.

A deficiency of vitamin C is in all probability more widespread than a deficiency of any other vitamin and much controversy still surrounds the significance of low tissue reserves of this vitamin in congenital defects, in wound healing and in malignant conditions. Other uncertainties relate to the possible role of megadose vitamin C therapy in the maintenance of health and to the relationship between smoking and ascorbic acid status. These questions are discussed by Dr Schorah.

Apart from primary vitamin deficiencies associated with poor diets — including diets consisting mainly of alcoholic drinks, a problem high-lighted by Dr Kelleher — secondary vitamin deficiency conditions induced by drugs present ever increasing problems in the community and these are considered by Professor Dickerson.

A secondary deficiency of vitamins also occurs in various malabsorption syndromes: folic acid and the fat-soluble vitamins are the ones most likely to be involved under these circumstances. Folic acid deficiency is, however, more common in pregnancy, expressing itself as a megaloblastic anaemia and this vitamin is considered in detail by Dr Chanarin.

Vitamin A is discussed by Dr Pitt and whereas a deficiency of this vitamin is widespread in South America, Indonesia and Southern India, often as part of a complex of nutritional deficiencies, there is little evidence of a lack of this vitamin in Britain.

There has been a great deal of nonsense written about vitamin E and its role in human nutrition and the most exaggerated claims have been made for its curative properties for a most diverse collection of diseases. In this context, it is refreshing to read Professor Losowsky's paper which sets the record straight. His conclusions are that vitamin E deficiency certainly occurs in the infant, particularly if pre-term, and that this deficiency must be treated. In adult men and women, however, deficiency occurs almost only in association with fat malabsorption syndromes and the consequences of this deficiency are minor.

Vitamin D deficiency is a major public health problem, particularly in cities where there is a large population of Asian immigrants. It may affect pregnant women and their unborn babies, infants, toddlers, adolescents and the elderly, causing rickets in infants and children and osteomalacia in adults. Professor Arneil discusses these problems and the inadequate response of the health services to them in a most forceful and challenging article.

The general view among nutritionists in the past has been that, although people are able to synthesize vitamin D in their skin by the action of ultraviolet light from the sun, the diet provides the major source of this vitamin. This view is now being questioned and Dr Lawson summarizes an impressive body of evidence indicating that, for most people, sunlight makes a greater contribution than food to their supplies of vitamin D. It would seem, therefore, that low exposure to sunlight is the main reason for the existence of vitamin D deficiency in vulnerable groups of the community although the main hope for preventing and curing this deficiency in these groups is still by way of the diet or by supplements of the vitamin.

We are agreed that vitamin deficiencies cause particular diseases but what are the effects of other disease states on the vitamin status of the patient? Dr Kelleher explores these relationships with particular reference to diseases of the liver, malabsorption syndromes and cancer.

Much of the food we consume has undergone some form of processing, whether it be milling, freezing, canning, drying, or even more sophisticated forms of processing, and an account of the effects of some of these processes and of domestic cooking on the vitamin content of foods is given by Professor Bender. Another chapter by the same author is concerned with the enrichment of manufactured foods with vitamins.

Analysts have always regarded vitamins as difficult substances to determine in foods. There are a number of different reasons for this: some are unstable, some that are unavailable to man are rendered available by the laboratory procedures used to extract them, some are present in a number of different forms varying in their biological activity, and for many vitamins the analytical techniques that are used are long and laborious. These problems are discussed by Dr Wiggins and in the two chapters that he has contributed to this book he describes some of the recent advances that have been made in analysing foods for both fat-soluble and water-soluble vitamins, emphasising in particular the use of the powerful new analytical tool of high performance liquid chromatography (HPLC).

This book thus provides a comprehensive and up-to-date account of the present state of knowledge of all aspects of vitamins with particular relation to their effects on health. It does in fact amply fulfil the promise of its title and we should all be grateful to the Kellog Company of Great Britain Ltd for commissioning a book which will be of immense value to all who have a professional interest in human nutrition.

T.G. TAYLOR
SOUTHAMPTON

1
Man's needs for vitamins – a need for review?

J. W. T. DICKERSON

INTRODUCTION

uphold

The amounts of energy, nitrogen and minerals needed to maintain the metabolic processes of the human body in a steady state can be determined with some degree of precision. The amounts of these nutrients supplied by the diet and the amounts leaving the body in the various forms and by the various routes can be measured and the difference between intake and output, expressed as the 'balance', is the change in the body content of the particular constituent. The body's 'needs' in terms of the amounts necessary to maintain the body of an adult 'in balance' and to provide the extra amounts needed for growth, or during pregnancy and lactation, can be assessed in a similar way.

The determination of man's needs for vitamins presents problems of a different kind. The needs for some of them, e.g. thiamin, riboflavin and nicotinic acid, because of their important role in energy metabolism, are usually related to the intake of energy and expressed per 1000 kcal or per MJ. For other vitamins, and notably for vitamin C, reliance is usually made upon the determination of the amount required to prevent a deficiency disease with additional allowance being made, where necessary, for urinary losses. These methods of determination are crude and the endpoints imprecise. The question that we must ask, and which is posed by my title, is whether estimates of needs based upon them are the best that we can get, whether our endpoint – the relief of a deficiency disease – should not be reconsidered, and if reconsideration is desirable, whether there are any presently known facts or trends that should be taken into account.

NOMENCLATURE

There is some confusion of nomenclature in this topic. As Wretlind and his colleagues[1] have pointed out 'dietary requirements', 'recommended intakes'

1

and 'allowances' are sometimes used interchangeably. Our 'needs' for a nutrient are the physiological requirements and it is very difficult to determine this because of the variety of criteria that can be used. Prevention of disease is not the same as the promotion of health and efficiency. In the context of the latter it may well be that 'needs' approximate to 'recommended intakes'. Waterlow[2] has emphasized that our task as nutritionists is to find out more about our needs.

PRESENT RECOMMENDED INTAKES

The recommended intakes of vitamins, like those of other nutrients, vary in different parts of the world (Table 1). These variations are due to a number of factors, including differences in the interpretation of investigations to determine requirements, in the amounts that are to be added to cover individual variations, and in the interpretation of the safety margin. Differences between countries are greater for some vitamins, e.g. vitamin C, than for others, e.g. niacin, and values for the USSR are generally higher than those for other countries. With respect to our own values we need to enquire as to whether we are satisfied with the way in which they were obtained and whether there have been any changes in the environment, or in the population, which make it worthwhile to review them.

Table 1 Recommended intakes in some European countries and in the United States

	Denmark	Germany (Fed. Rep.)	Nether-lands	Sweden	UK 1969	US RDA 1974	Germany (Dem. Rep.)	USSR
Vitamin A (μg)*	1000	900	850	900	750	1000	600	1500
Thiamin (mg)	1.4	1.6	1.2	1.4	1.2	1.4	1.6	1.8
Riboflavin (mg)	1.6	2.0	1.7	1.7	1.7	1.6	1.6	2.4
Niacin equiv. (mg)	18	9–15	—	18	18	18	18	20
Vitamin C (mg)	45	75	50	60	30	45	70	75
Vitamin D (μg)	2.5	2.5	—	10	2.5	—	—	12.5

*Retinol equivalents

We are not here considering the desirability, or otherwise, of using 'mega' or pharmacological doses of vitamins, or even of taking vitamin supplements, but whether there are grounds for reviewing the present recommended intakes. Consideration of the nature of our population, the changes that have taken place in the environment during the past thirty years, and identifiable future trends, suggests that there are three matters that may bear upon the problem. These are the interaction of vitamins with drugs and with the process of aging and the development of degenerative disease and any evidence that there may be that higher intakes improve health and efficiency.

DRUG–VITAMIN INTERACTIONS

It may seem strange to suggest that drugs should enter into a consideration of the physiological requirements for nutrients. However, the ready availability of drugs prescribed within the National Health Service combined with the ease with which many pharmaceutical preparations can be obtained over the counter without a prescription – self-medication – has led to the United Kingdom becoming a nation of drug-takers[4]. Drugs may affect the absorption and utilization of vitamins. They may also increase the needs for certain vitamins by virtue of the fact that they induce the mixed function oxidase system which exists mainly in the liver and which is responsible for converting drugs into forms in which they can be excreted in the bile or urine. Certain of these drug–vitamin interactions are known and cannot be ignored. In the assessment of their importance the production of a deficiency disease cannot be used as a criterion of deficiency. Blood and tissue depletion with breakdown of metabolism precede clinical disease. Furthermore, the widespread consumption of certain drugs, often self-prescribed, makes it impossible to rely upon advice for each individual by a doctor, even if the latter were aware of the interactions. It would therefore seem that for certain vitamins possible interaction with drugs should be considered in relation to needs on a community basis. This may be true for vitamin C and folic acid and perhaps also for thiamin.

Ascorbic acid (vitamin C)

Consideration of factors that are likely to affect the needs for ascorbic acid should be viewed against the background of the experimental basis for our recommended intake (Table 1). This value is based on a study[5] carried out in Sheffield between October 1944 and February 1946. Twenty volunteers, all men between the ages of 17½ and 34½ years, were divided into three groups. All received the same basic diet containing no vitamin C, but three men received supplements of 70 mg daily, seven received 10 mg daily and the remaining ten received no supplement. All the men that received only the basic diet developed scurvy and this was quickly relieved by 10 mg of vitamin C. One of these men was admitted to hospital as an acute emergency which could have been due to a scorbutic haemorrhage.

On the basis of this study it was concluded that the League of Nations figure of 30 mg vitamin C daily was a safe allowance. Though not producing

3

conclusive results, the tests of physical fatigue left some doubt as to whether 10 mg was an optimum dose and there were small differences in favour of 70 mg. The Special Report of this study goes on to say 'Any assessment is, in the present state of knowledge a matter of judgement and must be regarded as provisional'. It further adds 'When the figure of 30 mg is used, for whatever purpose, it should be borne in mind how it was assessed . . . Intakes much below the recommended figure . . . are not necessarily detrimental to health'.

Aspirin is one of the most commonly used drugs. Some years ago it was estimated that the annual consumption was 4000 million tablets and there were over 300 formulations containing aspirin on the market. There is no evidence that aspirin causes scurvy but the therapeutic dose has been shown[7] to prevent the uptake of vitamin C by leukocytes.

Oral contraceptive agents (OCA) cause a reduction in leukocyte vitamin C levels. The mechanism of this effect, like that on folic acid, is probably via oestrogen-induced cortisol secretion which induces a rise in mixed function oxidase activity in the liver. These enzymes require both vitamin C and folic acid as cofactors. However, there is also some evidence[8] that OCAs accelerate the breakdown of vitamin C and cause a change in tissue distribution of the vitamin. Smoking also reduces blood vitamin C levels. It is thus possible to visualize in some women a situation developing in which a state of subclinical vitamin C deficiency affects brain function[9] in subtle ways that are only apparent on careful assessment, but which nevertheless represent a reduction in the ability to work efficiently. OCAs also reduce the blood levels of other vitamins[4], notably pyridoxine.

Folic acid

Not all countries quote a recommended intake for folic acid. Our need for this vitamin is usually associated with blood cell formation. However, the vitamin plays an important role in DNA synthesis wherever that occurs, and the effects of subclinical deficiencies on cell multiplication in the intestinal mucosa and consequently upon the absorption of other nutrients remains to be elucidated. Subclinical deficiencies of this vitamin seem to be amongst the most widespread of all deficiencies in the UK. These are exacerbated by long-term therapy with anticonvulsants[10] and probably by long-term treatment with any drug which induces the drug-metabolizing enzymes. Low blood levels of folate are also induced by the consumption of alcohol which impairs the absorption of the vitamin[11].

Thiamin

Alcohol is a drug which is widely consumed. It is also a source of energy. Over-consumption of alcohol results in a reduction of intake, and of absorption, of thiamin whilst at the same time increasing its utilization. Evidence of frank deficiency (i.e. beri-beri) is rare in the UK and the Wernicke–Korsakoff

4

syndrome occurs in only a few alcoholics. However, biochemical evidence of thiamin deficiency is relatively common amongst the elderly in whom it may be associated with confusion. The widespread consumption of antacids, particularly amongst this group but also amongst younger people who suffer from dyspepsia, may lead to impaired absorption of the vitamin.

Vitamins and ageing

'Prevention' is now everybody's business[12] and this is nowhere more so than with respect to ageing. The structure of our population is changing and is likely to continue to do so. Whilst the life expectancy of a person reaching the age of 40 is only a little longer than it was a century ago many more people are, in fact, surviving to old age. The health-care demands that this increasing number of old people makes on the health service and community will depend on the extent of degenerative and other disease in this population. In terms of preventive medicine it then seems prudent to examine carefully those factors that may affect the rate of ageing or the increase of degeneration. Our goal is healthy elderly persons, rather than simply more of them.

Vitamins affect ageing but our question is whether there is a need to re-examine our recommended intakes in order to endeavour to retard the process.

Of all the degenerative disorders associated with ageing and the elderly, arteriosclerosis should be carefully considered, for in the heart, peripheral circulation and brain this may be responsible for much ill-health, discomfort and loss of function. A lifetime approach to this problem seems prudent[13] and consumption of a diet with an increased ratio of polyunsaturated to saturated fat is now widespread. Vitamin E, a naturally occurring antioxidant, is needed to prevent peroxidation of polyunsaturated fatty acids. In view of its important role in connection with these fatty acids, it would seem reasonable to consider a recommended intake of vitamin E. Just how much is needed however, seems to be a matter of debate and detailed consideration led Horwitt[14] to suggest a range of 10–30 mg d-α-tocopherol per day.

Further support for considering vitamin E comes from its role in preventing peroxidation in relation to another phenomenon associated with ageing. The free-radical hypothesis, suggested to account for ageing processes such as the accumulation of the age-pigment lipofuscin in the heart and brain, suggests that ageing is caused by reactions involving highly reactive 'free radicals'. These reactions involve the peroxidation of lipid and, in the case of lipofuscin, the combination of lipid peroxides with protein. Antioxidants, such as α-tocopherol (vitamin E), combine with the free radicals and hence prevent the peroxidation of lipid.

Studies in experimental animals[15][16] have shown that addition of α-tocopherol to the diet in excess of the normally-accepted amounts increases survival rate and presumably delays ageing. There was also some evidence of a reduction in lipofuscin deposition in the brains of the survivors[17].

There have been a number of suggestions[18] that a deficiency of vitamin C in the diet contributes to arteriosclerosis. These suggestions are largely based on the effect of vitamin C on plasma cholesterol levels and on the metabolism of cholesterol. More recent studies[19] suggest that high density lipoproteins (HDL) may have a protective effect in decreasing atherosclerosis and ischaemic heart disease. In elderly men the plasma level of vitamin C has been reported to be significantly correlated with the level of HDL cholesterol[20] although the authors were careful to point out that this did not necessarily mean that the relationship was causative. In a recent study we[21] have been giving 1 g of ascorbic acid to elderly patients with atherosclerotic heart disease and have found that it increased HDL cholesterol levels in six weeks. Whether an intake of 100 mg per day given over a longer time would have the same effect is not known.

Blood levels of vitamin C fall with advancing age even in the absence of dietary deficiency[22], due to impaired absorption or increased utilization, thus introducing a further factor to be considered in connection with our 'need' of this vitamin.

EFFECTS OF HIGHER THAN RECOMMENDED INTAKES

Few controlled studies have been made of the effects of increasing the intake of vitamins, other than those involving mega doses. In this connection, there would seem to be a need for more research into refinements of criteria to be used – consideration of improvements to health rather than the prevention of disease. Such studies are, however, difficult to perform, and the range of response considerable. In one such controlled study[23], housewives were given an additional 240 mg of vitamin C daily in a fruit drink and there was some evidence that this resulted in increased ability to concentrate and an increased sense of well-being.

CONCLUSIONS

This brief review suggests that we are in a position to provide only tentative answers to the problem of the needs of man for vitamins in amounts other than those which, with an additional safety margin, protect against deficiency diseases. However, there would seem to be sufficient indication that this conclusion leaves no room for complacency. Up till now it is doubtful if questions have been asked in the right way to elucidate right answers to the problem. We return again to the crudeness of our tools and the necessity to re-examine them and that in the light of our goal – the promotion of health.

As far as individual vitamins are concerned, evidence is increasingly in favour of a review of the recommended intake of vitamin C. For other vitamins it may be that it would be sufficient to ensure that at least those amounts presently recommended are consumed by all the population.

Research on this problem should concentrate on determining both short and long-term physiological effects of higher intakes particularly in relation to ageing.

References

1. Wretlind, A., Hejda, S., Isaksson, B., Kubler, W., Truswell, A. S. and Vivenco, F. (1977). *Nutr. Metab.*, **21**, 244
2. Waterlow, J. (1977). Contribution to General Discussion. Round table on comparison of dietary recommendations in different European countries. *Nutr. Metab.*, **21**, 233
3. Second European Nutrition Conference, Munich 1976. (1977). Tables of recommended nutrient intakes in different European countries. *Nutr. Metab.*, **21**, 251
4. Dickerson, J. W. T. (1978). Some adverse effects of drugs on nutrition. *J. R. Soc. Health*, **98**, 261
5. Bartley, W., Krebs, H. A. and O'Brien, J. R. P. (1953). Vitamin C requirement of human adults. *Med. Res. Council*, Spec. Rep. Series No. 280
6. Wade, O. L. (1970). *Adverse Reactions to Drugs.* p. 85. (London: Heinemann Medical)
7. Loh, H. S., Watters, K. and Wilson, C. W. M. (1973). The effects of aspirin on the metabolic availability of ascorbic acid in human beings. *J. Clin. Pharmacol.*, **13**, 480
8. Rivers, J. M. (1975). Oral contraceptives and ascorbic acid. *Am. J. Clin. Nutr.*, **28**, 550
9. Kinsman, R. A. and Hood, J. (1971). Some behavioural effects of ascorbic acid deficiency. *Am. J. Clin. Nutr.*, **24**, 455
10. Labadorios, D., Obuwa, G., Lucas, E. G., Dickerson, J. W. T. and Parke, D. V. (1978). The effects of chronic drug administration on hepatic enzyme induction and folate metabolism. *Br. J. Pharmacol.*, **5**, 167
11. Wu, A., Chanarin, I., Slavin, G. and Levi, A. J. (1975). Folate deficiency in the alcoholic — its relationship to clinical and haematological abnormalities, liver disease and folate stores. *Br. J. Haematol.*, **29**, 469
12. Department of Health and Social Security. (1976). *Prevention and Health: Everybody's Business.* (London: HMSO)
13. Lloyd, J. and Wolff, O. H. (1969). A paediatric approach to the prevention of atherosclerosis. *J. Atheroscler. Res.*, **10**, 135
14. Horwitt, M. K. (1974). Status of human requirements for vitamin E. *Am. J. Clin. Nutr.*, **27**, 1182
15. Harman, D. (1968). Free radical theory of ageing: effect of free radical reaction inhibitors on the mortality rate of male LAF mice. *J. Geront.*, **23**, 476
16. Rudra, D. N. (1976). The effect of some antioxidants in the process of ageing. PhD Thesis. University of Surrey
17. Rudra, D. N., Dickerson, J. W. T., Walker, R. and Chayen, J. (1975). The effect of some antioxidants on lipofuscin accumulation in rat brain. *Proc. Nutr. Soc.*, **34**, 122A
18. Ginter, E. (1974). Vitamin C in lipid metabolism and atherosclerosis. In: G. G. Birch and K. Parker (eds.) *Vitamin C,* pp. 179-198. (London: Applied Science Publ.)
19. Miller, N. E. and Thelle, D. S. (1977). The Tromsø heart study. High density lipoprotein and coronary disease. A prospective case-control study. *Lancet*, **i**, 965
20. Bates, C. J., Mondal, A. R. and Cole, T. J. (1977). HDL cholesterol and vitamin-C status. *Lancet*, **ii**, 611
21. Horsey, J., Livesey, B. and Dickerson, J. W. T. (1979). Aged patients and ischaemic heart disease: effects of ascorbic acid on lipoproteins. (In preparation)
22. Burr, M. L., Sweetnam, P. M., Hurley, R. J. and Powell, G. H. (1974). Effects of age and intakes on plasma ascorbic acid levels. *Lancet*, **i**, 163
23. Canter, S. Personal communication

2

Developments in the determination of water-soluble vitamins in food

R. A. WIGGINS

Recent developments in the analytical methods for the determination of water-soluble vitamins have taken two main courses:

1. Automation of standard or well-established methods to cope with increasing demand for analytical data generated by nutritional survey programmes and food quality control.
2. Application of new analytical techniques to improve speed and specificity of analysis.

A well-established method for the determination of the B vitamins is microbiological assay and over the past few years the procedure used for the microbiological assay[1] of B vitamins at the Laboratory of the Government Chemist (LGC) has been automated[2] to improve sample throughput as well as to relieve much of the tedium of the manual procedure. Two automated modules have been constructed; one prepares diluted extracts and the second automatically measures the growth of the micro-organism. After conventional extraction[1] the food extracts and standard solutions of the vitamins to be assayed are loaded onto a carousel of the first module. A probe then sucks up these solutions in turn and dispenses accurate amounts (100, 200, 300 and 400 μl) into assay tubes contained in a rack. Nutrient medium is then automatically dispensed into the assay tubes. The rack of assay tubes is sterilized, inoculated manually and incubated, usually for about 20 h. The vitamin-dependent growth of the micro-organism is then measured turbidimetrically in the second automated module (Figure 1). This module automatically positions the assay tubes under a pneumatic probe which sucks up the contents of the tube, after mixing with a stream of air, into a flow-cell of a colorimeter. The turbidity of the solutions is measured and the data is transmitted to the in-house computer for analysis.

9

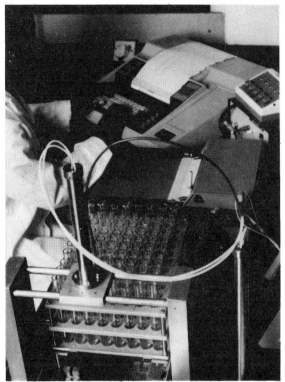

Figure 1 The second automated module (see text)

Although this batch automation does increase the throughput of samples and reduces the possibility of human error by automation of several of the tedious manual operations the time of analysis per sample is still approximately two days. This period of time can be inconvenient and this has prompted recent publications which indicate that it is possible to reduce the time of microbiological assay for some B vitamins. Japanese workers[3][4] have reduced the incubation time for the assay of thiamin and nicotinic acid to 6 h and 1 h respectively by inoculation with high concentrations of lactobacilli followed by electrochemical estimation of vitamin-dependent growth. The incubation time for nicotinic acid was reduced to 1 h by inoculation with high concentrations (10 mg wet cells ml^{-1}) of *Lactobacillus plantarum* immobilized in agar gel[4]. In this assay the growth of the lactobacilli was monitored by measuring the increase in hydrogen ion concentration with a glass electrode; for the thiamin assay a cell with platinum anode and silver peroxide cathode was used[3].

The incubation time for the determination of folate using *Lactobacillus casei* has been shortened by colorimetric measurement of the formazan formed by bacterial reduction of 2,3,5-triphenyl tetrazolium chloride (TTC). This procedure has been used in the design of a continuous flow assay for folate in serum[5]. Assay medium, inoculum (from a continuous culture) and sample extract are mixed with a proportioning pump and the mixture incubated for 3 h by passage through a long length of polythene tube maintained at 37 °C in a water bath. A solution of TTC is then directed into the flow system which is then incubated for a further 2 h at 37 °C. The growth is estimated by measurement in a flow cell of formazan absorbance at 500 nm.

Continuous flow analysis has also been used to automate the chemical analysis of thiamin[6-8], riboflavin[7 8], niacin[9] and vitamin C[10]. These automated analyses are based on AOAC procedures[11] and in most of the automation studies comparison of results determined by manual and automated procedures show overall good agreement. Usually the initial extraction of the vitamin from the food has not been automated: however, in most of the studies the extraction method has been examined and where possible shortened. As well as significantly improving the throughput of samples, automation of these chemical methods has the added advantage of increasing precision as a result of careful control of the addition and mixing of reagents to sample and standard. This increase in precision is particularly noticeable in the thiamin analysis where any slight variation in technique between sample and standard in the formation of thiochrome can result in significant error[6].

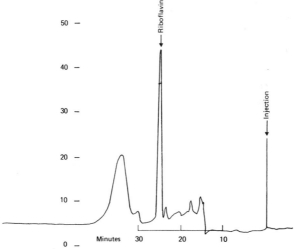

Figure 2 Determination of riboflavin in wholemeal flour by HPLC. Column – 5 micron Spherisorb ODS (25 cm × 4 mm); mobile phase – 30% methanol/70% water buffered to pH 5.8 with citrate; detector – fluorescence excitation 449 nm, emission 520 nm; sample loading – approx. 200 ng

The chemical analysis of thiamin and riboflavin in foods have been considerably simplified by the application of HPLC. Riboflavin can be separated in a matter of minutes on either an HPLC column of 10 micron silica[12] or on a column of 5 micron C18-reverse phase packing[13]. The latter column is eluted with 30% methanol/70% water buffered at pH 5.8 and the silica column with 0.1 M sodium acetate at pH 4.6. These separations have been applied to food extracts and the riboflavin detected fluorometrically (Figure 2). The fluorescence intensity of riboflavin varies with pH so it is important to buffer mobile phases, particularly when the riboflavin is calculated by comparing peak heights with those of standards. Care must also be taken in preparation of food extracts for HPLC analysis of riboflavin. Normally, extracts for riboflavin analysis are prepared by acid hydrolysis which converts the flavin adenine dinucleotide (FAD) present in foods to flavin mononucleotide (riboflavin-5' phosphate) (FMN) and riboflavin. The hydrolysed extracts are then incubated with an enzyme to convert the FMN to free riboflavin. An enzyme often used for this stage is takadiastase. However, recent investigations at the LGC have shown that several commercially available forms of takadiastase failed to convert quantitatively FMN to riboflavin even after several hours incubation. For this reason acid phosphatase is now used at the LGC to convert FMN to riboflavin for subsequent HPLC analysis. This enzyme produces quantitative conversion in 30 min at 44°C in citrate buffer at pH 5.6[13].

Columns of microparticulate silica with 0.1 M phosphate buffer (pH 6.8) plus 10% ethanol as mobile phase have been used to separate thiamin from food extracts[14]. The thiamin was detected by mixing column eluant with potassium hexacyanoferrate (III) to convert thiamin to thiochrome which was then measured fluorometrically.

HPLC has been applied to the separation of the three compounds that constitute vitamin B_6 and also to the separation of folates. Pyridoxol, pyridoxamine and pyridoxal have been separated on columns of the micro-

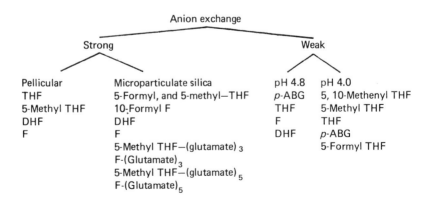

' **Figure 3** HPLC separation of folic acid derivatives

particulate cation exchange resin, Aminex A5[15]. Phosphate buffers of pH 4.3, 5.1 and 5.8 were used to separate pyridoxol, pyridoxamine and pyridoxal which were detected colorimetrically by mixing column eluant with a diazide in a continuous flow system. The diazide couples with the aromatic nuclei of the vitamins to give an orange product. Preliminary purification of extracts on a column of Amberlite ion exchange was necessary in order to remove brown pigments formed during acid hydrolysis. These Maillard browning pigments can also cause problems in the fluorometric analysis of vitamin B_6, particularly if the natural fluorescence is enhanced by conversion of the three forms to the 4-pyridoxic acid lactone, as the brown pigments react also with cyanide used to prepare lactone, to form highly fluorescent products. A fluorometric method for the analysis of vitamin B_6 in foods based on the formation of the pyridoxic acid lactone has recently been published[16]. In this method the brown pigments were removed and the three forms of vitamin B_6 separated on columns of Dowex AG-50W-A8 ion exchange resin. The fractions containing pyridoxol and pyridoxamine were reacted with manganese dioxide and sodium glyoxylate respectively to form pyridoxal and the three fractions of pyridoxal converted to the pyridoxic acid lactone by reaction with potassium cyanide.

Folates occur in food as a variety of compounds in which the basic folic acid (pteroylglutamic acid) differs in the state of oxidation of the pteridine ring, the presence or absence of substitution at the 5- of 10 nitrogen, and the number of glutamic acid residues. HPLC[17-20] has been applied to the separation of the individual folates and it would appear from published separations that before very long a routine analysis of food folate could also include a profile of the individual folates present providing that standards are available for calibration. Correlation of folate content determined by HPLC and microbiological assay will probably cause problems because of the variation in response of the micro-organisms used for assay to the various forms of folate present in a food extract. Figure 3 summarizes the published separations. It was found that the retention of folates on the pellicular strong anion exchange resin was dependent on the pteridine part of the molecule[17]. Consequently the polyglutamates eluted with corresponding monoglutamates. This separation has been applied to several foodstuffs. Conversely the separation on the siliceous strong anion exchanger[18] appeared to depend on the interaction between folate carboxyl groups and quaternary ammonium groups of resin. This system could therefore be used to separate the polyglutamates.

The ion-exchange HPLC systems Aminex A14/0.1 M β-alanine–NaCl–0.01 M Na_2SO_3, pH 4.5[21] and Zipax SAX/0.07 M acetate pH 4.8[22] have been used to determine ascorbic acid in foods. The Zipax SAX system was used with an electrochemical detector. Reverse phase ion pair chromatography[23 24] has also been used to determine ascorbic acid (Figure 4)[24]. In this technique a neutral ion pair is formed with ionized ascorbic acid by adding a quaternary ammonium salt to the methanol/water mobile phase.

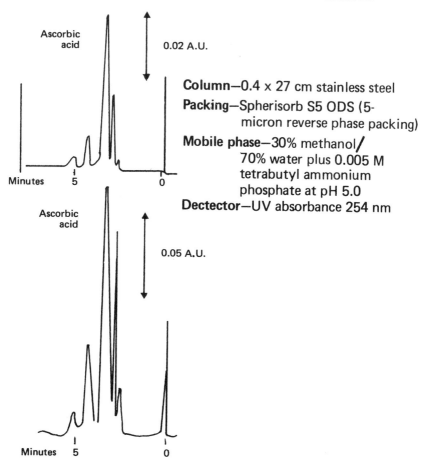

Column—0.4 x 27 cm stainless steel
Packing—Spherisorb S5 ODS (5-
micron reverse phase packing)
Mobile phase—30% methanol/
70% water plus 0.005 M
tetrabutyl ammonium
phosphate at pH 5.0
Dectector—UV absorbance 254 nm

Figure 4   HPLC determination of ascorbic acid in ham

The neutral species can then be successfully analysed on reverse phase packings which is an advantage in food analysis as these packings are less easily irreversibly contaminated than ion exchange or silica packings. The ion exchange separation on Zipax SAX[22] and the reverse phase separations[23] have been compared with titrimetric methods for determination of ascorbic acid. The trend was for HPLC methods to give lower results than the titrimetric method which is attributed to the increased selectivity of the HPLC method. These HPLC systems are not suitable for measurement of dehydro-ascorbic acid directly. Any contribution of dehydroascorbic acid to vitamin C activity will be missed unless it is reduced to ascorbic acid before analysis. A current project at the LGC is concerned with developing an HPLC procedure that will measure dehydroascorbic acid directly. This project has involved investigation of the possibility of preparing ultraviolet absorbing derivatives of dehydroascorbic acid.

Competitive protein binding methods, previously applied to the analysis of biological fluids, have been applied to the analysis of folate and vitamin B_{12} in foods. They have at least equal sensitivity to the microbiological assay methods and are quicker. Correlation between competitive protein binding assay methods and microbiological assay are, however, often not very good. In the case of folate analysis, problems arise because of the different affinities of protein and micro-organisms to the individual folates present in extracts[25]. This difference in response to various forms of a vitamin is possibly also the reason for the poor agreement found with some foods in a recently reported comparative study of the determination of vitamin B_{12} using a commercially available radioassay kit and microbiological assay with *L. leichmanii*[26].

References

1. Bell, J. G. (1974). Microbiological assay of vitamins of the B group in foodstuffs. *Lab. Practice*, **23**, 235
2. Stockwell, P. B. (1978). Automatic methods of food analysis. In R. D. King (ed.). *Developments in Food Analysis Techniques*, p. 255. (London: Applied Science Publishers).
3. Matsunaga, T., Karube, I. and Suzuki, S. (1978). Electrochemical microbioassay of vitamin B_1. *Anal. Chim. Acta*, **98**, 25
4. Matsunaga, T., Karube, I. and Suzuki, S. (1978). Rapid determination of nicotinic acid by immobilized *Lactobacillus arabinosus*. *Anal. Chim. Acta*, **99**, 233
5. Tennant, G. B. (1977). Continuous-flow automation of the *Lactobacillus casei* serum folate assay. *J. Clin. Pathol*, **30**, 1168
6. Kirk, J. R. (1974). Automated method for the analysis of thiamine in milk, with applications to other selected foods. *J. Assoc. Offic. Analyt. Chemists*, **57**, 1081
7. Pelletier, O. and Madere, R. (1975). Comparison of automated and manual methods for determining thiamine and riboflavine in foods. *J. Food Sci.*, **40**, 374
8. Pelletier, O. and Madere, R. (1977). Automated determination of thiamine and riboflavin in various foods. *J. Assoc. Offic. Analyt. Chemists*, **60**, 140
9. Egberg, D. C., Potter, R. H. and Honold. G. R. (1974). The semiautomated determination of niacin and niacinamide in food products. *J. Agr. Food Chem.*, **22**, 323
10. Egberg, D. C., Potter, R. H. and Heroff. J. C. (1977). Semiautomated method for the fluorometric determination of total vitamin C in foodstuffs *J. Assoc. Offic. Analyt. Chemists*, **60**, 126
11. *Official Methods of Analysis of the Association of Official Analytical Chemists*. (1975). 12th edition, Section 43. (Washington D.C.: Assoc. Offic. Analyt. Chemists).
12. Richardson, P. J., Favell, D. J., Gidley, G. C. and Jones, A. D. (1978). A critical comparison of the determination of vitamin B_2 in foods by a new high-performance liquid chromatographic method and the 'standard' microbiological approach. *Proc. Analyt. Div. Chem. Soc.*, 53
13. Lumley, I. and Wiggins, R. A. Unpublished results
14. Van de Weerdhof, T., Wiersum, M. L. and Reissenweber, H. (1973). Application of liquid chromatography in food analysis. *J. Chromatogr.*, **83**, 455
15. Yasumoto, K., Tadera, K., Tsuji, H. and Mitsuda, H. (1975). Semi-automated system for analysis of vitamin B_6 complex by ion-exchange column chromatography. *J. Nutr. Sci. Vitaminol.* **21**, 117
16. Gregory, J. F. and Kirk, J. R. (1977). Improved chromatographic separation and fluorometric determination of vitamin B_6 compounds in foods. *J. Food Sci.*, **42**, 1073
17. Clifford, C. K. and Clifford, A. J. (1977). High pressure liquid chromatographic analysis of food in folates. *J. Assoc. Office. Analyt. Chemists*, **60**, 1248
18. Stout, R. W., Cashmore, R. A., Coward, J. K., Horvath, G. G. and Bertino, J. R. (1976).

Separation of substituted pteroylmonoglutamates and pteroyl oligo – L-glutamates by high pressure liquid chromatography. *Anal. Biochem.*, **71**, 119

19. Reed, L. S. and Archer, M. C. (1976). Separation of folic acid derivates by high performance liquid chromatography. *J. Chromatogr.*, **121**, 100

20. Branfman, A. R. and McCornish, M. J. (1978). Rapid separation of folic acid derivatives by paired-ion high performance liquid chromatography. *J. Chromatogr.*, **151**, 87

21. Floridi, A., Fini, C., Palmerini, C. A. and Rossi, A. (1976). Dosaggio cromatografico delle vitamine. I. Comportamento cromatografico delle vitamine idrosolubili su resine Aminex. *Riv. Sci. Tecn. Alim. Nutr. Um.*, **6**, 197

22. Pachla, L. A. and Kissinger, P. T. (1976). Determination of ascorbic acid in foodstuffs, pharmaceuticals and body fluids by liquid chromatography with electrochemical detection. *Anal. Chem.*, **48**, 364

23. Snood, S. P., Sartori, L. E., Wittner, D. P. and Haney, W. G. (1976). High-pressure liquid chromatographic determination or ascorbic acid in selected foods and multivitamin products. *Anal. Chem.*, **48**, 796

24. Lumley, I. (1979). The determination of ascorbic acid and dehydroascorbic acid by high performance liquid chromatography. MSc Thesis, Polytechnic of the South Bank, London

25. Ruddick, J. E., Vanderstoep, J. and Richards, J. F. (1978). Folate levels in food – A comparison of microbiological assay and radioassay methods for measuring folate. *J. Food Sci.*, **43**, 1238

26. Richardson. P. J., Favell, D. J., Gidley, G. C. and Jones, G. H. (1978). Application of Commercial radioassay test kit to the determination of vitamin B_{12} in food. *Analyst*, **103**, 865

16

3
Vitamin B$_1$ supply in industrialized countries

G. B. BRUBACHER

The diet of Western countries is characterized in several ways, two of these being a low intake of energy (due to steadily decreasing physical activity) and a relatively high intake of refined foods.

It is not unusual to have a diet made up as follows:-

Total energy intake (%)	Food
15	Sugar
5–10	White flour*, rice or products from refined starch or gelatine
20	Refined fats and oils
5–10	Alcoholic beverages

This means that with a daily diet of 10 MJ (2400 kcal) an average individual may eat about 90 g sugar, 30–60 g white flour, rice and so on and drink about half a litre of beer or two to three glasses of wine.

Under such conditions there are only 4.6–5.4 MJ (1100–1300 kcal) left in which all the necessary nutrients besides energy have to be incorporated. So it is difficult to provide a nutritionally adequate diet. One of the critical nutrients is certainly vitamin B$_1$ and if we assume that under optimal conditions at least 0.1 mg vitamin B$_1$/MJ (0.4 mg/1000 kcal) should be ingested, we can readily understand that under the described conditions only food with a nutrient density of 0.2 mg/MJ (0.8 mg/1000 kcal) and more contributes to the vitamin B$_1$ supply, since with each MJ rich in vitamin B$_1$ we also eat one MJ void of vitamins. From food composition tables we can calculate that only vegetables, oranges, potatoes, some cereals, pork and veal

---

* In the UK white flour is enriched with thiamin by law: therefore, white flour and white bread should not be considered as refined foods in this respect in the UK in contrast to countries such as West Germany, France and Switzerland.

Table 1   Thiamin content (vitamin B_1) in food

	Energy content/100 g		Thiamin density		
	kJ	kcal	mg/MJ	mg/1000 kcal	
Asparagus	88	21	2.05	8.6	
Spinach	63	15	1.60	6.7	
Wheat germ	1674	400	1.19	5.0	
Cauliflower	84	20	1.19	5.0	
Ox heart	536	128	0.98	4.1	
Fresh green peas	352	84	0.91	3.8	
Pig heart	511	122	0.91	3.8	
Pig liver	544	130	0.78	3.3	
Ox liver	544	130	0.78	3.3	
Kale	134	32	0.74	3.1	
String beans	155	37	0.65	2.7	
Pork	1130	270	0.60	2.5	
Tomatoes	80	19	0.60	2.5	
Oranges	176	42	0.57	2.4	
Brussel sprouts	218	52	0.50	2.1	High content
Carrots	147	35	0.48	2.0	
Common cabbage	100	24	0.48	2.0	
Smoked ham	1630	390	0.43	1.8	
Dried lentils	1420	340	0.36	1.5	
Orange juice	205	49	0.33	1.4	
White beans	1470	352	0.31	1.3	
Potatoes	377	90	0.26	1.1	
Wholemeal bread	1000	240	0.26	1.1	
Pickled cabbage	110	26	0.24	1.0	
Oats	1683	402	0.24	1.0	
Bacon	2220	530	0.22	0.9	
Veal	670	160	0.22	0.9	
Chanterelle	96	23	0.21	0.87	
Rice, whole	1500	360	0.19	0.8	
Strawberry	163	39	0.18	0.77	
Yolk of egg	1578	377	0.18	0.77	
Whole egg	678	162	0.18	0.74	
Rye-bread	963	230	0.17	0.70	
Whole milk	280	67	0.13	0.53	
Apple	218	52	0.12	0.52	
Liver-sausage	1880	450	0.11	0.47	
Walnuts	2724	651	0.11	0.46	
Cheese brie (50% fat)	527	126	0.10	0.40	
White of egg	226	54	0.10	0.40	
Vienna sausage (Frankfurter)	1105	264	0.09	0.38	
White bread	1088	260	0.08	0.33*	
Salami	2300	550	0.08	0.32	
Semolina	1548	370	0.08	0.32	Low content
Beef	1059	253	0.07	0.28	
Cheese (10% fat)	795	190	0.06	0.26	
Herring	1067	255	0.05	0.21	
Cottage cheese	398	95	0.05	0.21	
Polished rice	1548	370	0.04	0.16	

18

Table 1 (cont.)

	Energy content/100g		Thiamin density		
	kJ	kcal	mg/MJ	mg/1000 kcal	
Cheese (hard, 45% fat)	1736	415	0.03	0.12	⎤
Apple juice	197	47	0.03	0.12	⎬ Low content
Butter	3250	777	0.00	0.01	⎦

*When unfortified with thiamin

fulfil the above condition, whereas most fruits, eggs, some other cereals, milk and beef have less vitamin B$_1$ than the required 0.2 mg/MJ (see Table 1). This means that with the incorporation of one of these items in the diet at least one item with a content of more than 0.2 mg vitamin B$_1$/MJ should also be included to compensate for the low vitamin B$_1$ level of the other items. Moreover, it has to be taken into account that some foods rich in vitamin B$_1$ can lose up to 40% of their vitamin B$_1$ content or even more during the preparation of the meal. Thus with 50% of the energy void of vitamin B$_1$ the remaining food items not only have to be chosen carefully but they have to be prepared in such a way that cooking losses are negligible, as otherwise, an adequate supply of vitamin B$_1$ cannot be guaranteed.

Recommended dietary allowances are laid down so as to provide a reasonable safety margin. Now we shall consider what happens if these recommended dietary allowances are not fulfilled. From Figure 1 we can see that, depending on the intake of vitamin B$_1$, a series of vitamin B$_1$ depletion stages can be distinguished. With a high intake of vitamin B$_1$, resorption of vitamin B$_1$ is lowered and excretion of unaltered or metabolized vitamin B$_1$ is high. These two mechanisms give a certain guarantee that the body stores of vitamin B$_1$ do not fluctuate too much with the fluctuation of vitamin B$_1$ intake so long as this intake is in the high range, but with a lower intake of vitamin B$_1$ body stores decrease, which can be seen as the first stage of vitamin B$_1$ depletion by measuring vitamin B$_1$ excretion or vitamin B$_1$ concentration in blood plasma.

The next stage of an impaired vitamin B$_1$ supply consists of a diminished concentration of vitamin B$_1$ metabolites in blood and urine, the most common of which is represented by thiamin pyrophosphate. If the vitamin B$_1$ supply is even lower, we observe that the activity of vitamin B$_1$ dependent enzymes is reduced. But this does not necessarily mean that the metabolism of the body is disturbed. Most enzymes are produced in much greater amounts than are needed to maintain steady state concentration of the various compounds of intermediary metabolism. Only key enzymes, whose activity is, in general, regulated by feedback mechanisms, have to be constant in their activity. With

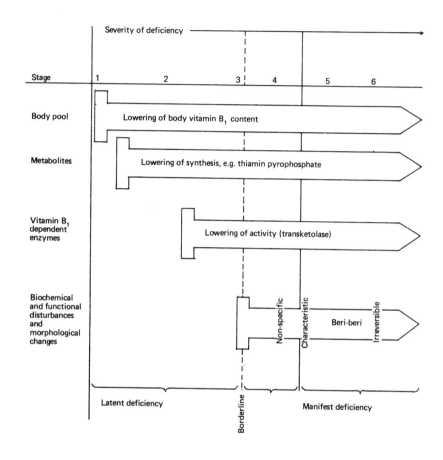

Figure 1 Stages of vitamin B_1 deficiency

vitamin B_1 we can measure the activity of the enzyme transketolase in red blood cells, an enzyme whose activity depends on the presence of thiamin pyrophosphate. Since it is not easy to standardize the assay of this enzyme because there is no suitable control preparation, the measurement in practice is made in a haemolysate of erythrocytes with and without the addition of thiamin pyrophosphate (TPP): enzyme activity in the presence of TPP is a measure of the concentration of the apoenzyme. We assume that the concentration of the apoenzyme is independent of the vitamin B_1 supply; this is more or less true in the region of a vitamin B_1 supply of 0.1 mg vitamin B_1/MJ and over. If we call this activity ETK (+TPP) and the activity measured direct without addition of ETK_0, the quotient

$$\alpha_{ETK} = \frac{ETK\,(+TPP)}{ETK_0}$$

is independent of standardization procedures and can be regarded as a measure of the vitamin B$_1$ supply status. The lower the vitamin B$_1$ supply, the higher is the value of α_{ETK}.

In the next stage, the activity of some vitamin B$_1$ dependent enzymes becomes rate-limiting and therefore, at this stage, metabolism becomes disturbed as can be seen by biochemical, morphological or functional changes. For example, at this stage, the concentration of pyruvic acid in blood is raised compared to control subjects with adequate intake of vitamin B$_1$, mainly after a glucose load and exercise. In population groups with a vitamin B$_1$ intake of 0.07 mg/MJ or less, the well known clinical signs of beri-beri may occasionally occur which becomes endemic at an intake level of about 0.05 mg/MJ or less.

We call the first three stages 'latent vitamin B$_1$ deficiency stages', whereas the transient area from the third to the fourth stage can be considered as a 'borderline' vitamin B$_1$ deficiency. Owing to the biological variation for a single individual no exact figure can be given for the cut-off points between adequate vitamin B$_1$ supply and latent vitamin B$_1$ deficiency, between latent vitamin B$_1$ deficiency and borderline vitamin deficiency and between borderline vitamin B$_1$ deficiency and manifest vitamin B$_1$ deficiency, but in epidemiological studies some guidelines have to be given which associate a given value of vitamin B$_1$ intake or vitamin B$_1$ excretion or α_{ETK} and so on, with a certain risk of the individual falling into one of these categories.

Body stores of vitamin B$_1$ are very low, and with an inadequate diet a stage of latent vitamin B$_1$ deficiency develops within a few days. To demonstrate this fact we chose 25 volunteers and allowed them to continue with their regular diet but asked them to avoid pork, vitaminized cereals or bread and vitaminized breakfast drinks. With these mild restrictions it was possible to see a significant fall in vitamin B$_1$ excretion after 14 days and at the same time a fall in transketolase activity or a rise in the α_{ETK} value. Whereas at the beginning of the experiment only about 15% of the volunteers were inadequately supplied with vitamin B$_1$, the figure was 30% at the end of the period as judged by thiamin excretion or transketolase values. By comparing the effect of the intake of a supplement of 1 mg vitamin B$_1$ and 2 mg vitamin B$_1$ per day, respectively, under the same dietary restrictions we were able to estimate the contribution of vitamin B$_1$ supply by pork, enriched cereals or bread and fortified breakfast drinks to be between 0.025 and 0.01 mg vitamin B$_1$/MJ. From this trial, it can be concluded that, under normal conditions, the diet of the volunteers contained no safety margin and that by dietary restrictions falling within the range of normal eating habits, the supply of vitamin B$_1$ may fall below the optimum amount, resulting in latent or even borderline vitamin B_1 deficiency appearing within a short time[1].

In an epidemiological study which we have conducted in Basle we have found that the frequency of persons with borderline vitamin B_1 deficiency, as judged by the transketolase test, is higher during the summer months than during winter. A possible explanation for this fact could be that in summer consumption of 'empty calories' such as ice cream and soft drinks is higher and therefore, less space is left for the intake of vitamin B_1 rich food. On the other hand, two items with a high nutritional density of vitamin B_1, oranges and pork, are mainly consumed during the cold season[2].

At the stages of latent vitamin B_1 deficiency, there is no immediate health danger since all physiological functions of the organism are still fulfilled. The danger lies mainly in the fact that, in cases where a new situation arises which requires higher intakes of vitamin B_1, the latent vitamin B_1 deficiency becomes of borderline or even manifest vitamin B_1 deficiency status. One of these situations is pregnancy, a situation where borderline vitamin B_1 deficiency is not uncommon due to a higher requirement of vitamin B_1[3]. Another situation is alcoholism where the absorption of vitamin B_1 may be reduced and therefore, a higher intake is needed. If no countermeasures are taken in this situation even manifest vitamin B_1 deficiency can be observed[4].

Since borderline vitamin B_1 deficiency is not uncommon as we have seen, for instance, in the epidemiological study cited above[2] (see also Ref. 5), we have to ask whether or not this stage has an impact on health or well-being. From the literature it can be derived that this impact certainly exists but it is not known to what extent this has consequences to the general public health. In this direction more research is needed. In Table 2 some possible influences of a borderline vitamin B_1 deficiency on the organism are listed.

Table 2 Manifestations of a borderline vitamin B_1 deficiency

Manifestation	Reference
Abnormal mental behaviour	6
Abnormal fetal development	3
Increased morbidity	9

In the study of Brozek[6] volunteers were kept for 161 days on a diet with a constant vitamin B_1 content: four volunteers received 0.19 mg thiamin/1000 kcal per day, four volunteers 0.31 mg thiamin/1000 kcal per day and two volunteers 0.55 mg thiamin/1000 kcal per day. After this time, all volunteers received a diet deprived of thiamin as long as ethical considerations permitted. Afterwards every subject received a supplement of 5 mg thiamin a day. As expected, the subjects of the first two groups could be considered to be in a latent vitamin B_1 deficiency state since it took, after complete deprivation of thiamin only 3–7 days before nausea, the first sign of a manifest vitamin B_1 deficiency, appeared in the group with 0.19 mg/1000 kcal vitamin B_1 intake and 7–15 days in the group of 0.31 mg/1000 kcal intake, whereas the two subjects with an intake of 0.55 mg/1000 kcal developed nausea only after 16 – 17 days. The group receiving only 0.19 mg thiamin/1000 kcal showed, after 161 days, a higher score for hypochondriasis, depression and hysteria in the Minnesota Multiphasic Personality Inventory Test, three personality changes which became highly significant in all three groups after total vitamin B_1 deprivation. Of course, the small number of subjects did not allow differentiation between the three groups by statistical means. Nevertheless, it is reasonable to assume that borderline vitamin B_1 deficiency is accompanied by a tendency to hypochondriasis, depression and hysteria. Whether or not this tendency has consequences in economic terms has to be investigated.

In the study of the Deutsche Forschungsgemeinschaft called 'outcome of pregnancy and development of child'[7] W. Kübler has investigated the correlation between vitamin B_1 status of the mother and the birthweight of the child. It was found that during the first trimester, about 11% of the pregnant women involved in this study were on borderline vitamin B_1 supply judged by the transketolase test and that the birthweight of the child was positively correlated to the vitamin B_1 status of the mother[2]. Since perinatal mortality increases as birthweight decreases, the association between vitamin B_1 status of the mother and birthweight of the child cannot be neglected from a public health point of view and it should be investigated whether or not a causal relationship exists between these two observed values and whether the low birthweight can be considered as a borderline vitamin B_1 deficiency symptom.

A third example is given by the observation of A. Lemoine[8], who measured the supply of vitamin B_1 in 656 healthy persons or patients with diabetes, liver diseases and gastrointestinal troubles. At the same time, he measured their vitamin B_1 status by means of the transketolase test and the number of non-specific symptoms such as skin or neuropsychological lesions which could also be caused by malnutrition. He found that there was a good correlation between the dietary and biochemical findings and that subjects with a lower vitamin B_1 intake or with an unacceptable vitamin B_1 status suffered in general from more non-specific symptoms than subjects with a high vitamin B_1 intake or with a good vitamin B_1 status. These results are shown in Figures 2 and 3, in which it can be seen that, for example, a higher percentage of subjects in the marginal group had more than two or three non-specific symptoms than in the adequate group, which means that a borderline vitamin B_1 deficiency status is associated with a higher degree of morbidity. Of course, these associations are no proof of a causal relationship but such a relationship is highly suggestive and needs further investigation.

In conclusion, it can be said that, in modern society, as a consequence of eating habits and lack of exercise, vitamin B_1 will be one of the nutrients which will become or may even already be a limiting factor in the nutrition of the

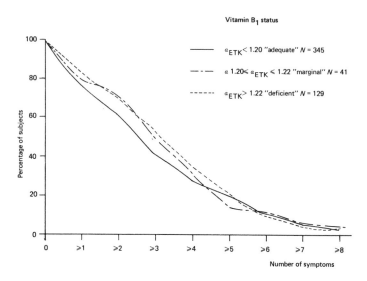

Figure 2 Correlation between number of non-specific symptoms and the activation coefficient for erythrocyte transketolase α_{ETK} (see p. 23)(Ref. 8)

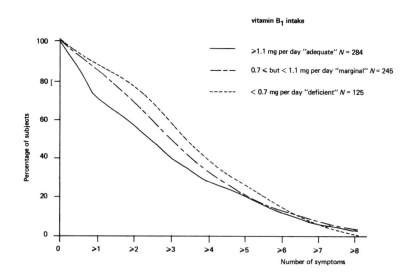

Figure 3 Correlation between number of non-specific symptoms and vitamin B_1 intake (Ref. 8)

general population. Although there is no proof that this lack of vitamin B_1 has untoward consequences from a public health point of view – and more research is needed in this respect – there is enough evidence to suspect that such a health problem exists and, since time is short, authorities should not wait until the proof is given but try to counteract by suitable measures the present development!

References

1. Brubacher, G., Haenel, A. and Ritzel, G. (1972). Zur Thiaminversorgung eines städtischen Kollektivs von Studenten. *Int. J. Vit. Nutr. Res.*, **42**, 451
2. Ritzel, G. (1974). Die Beeinflussung der Nährstoffversorgung durch äussere und innere Faktoren, aufgezeigt am Beispiel der Vitamine C und B_1. In: G. Brubacher and G. Ritzel (eds.). *Qualitätskriterien der Nahrung*, pp. 80–92. Int. Z. Vitaminforschung, Beiheft 14. (Bern, Stuttgart, Wien: Hans Huber Verlag)
3. Kübler, W. and Moch, K. J. (1975). Diskussionsvotum zum Vortrag K. Decker *et al.* In: G. Brubacher and G. Ritzel (eds.). *Zur Ernährungssituation der schweizerischen Bevölkerung*. pp. 233–241. (Bern, Stuttgart, Wien: Hans Huber Verlag)
4. Hell, D., Six, P. and Salkeld, R. (1976). Vitamin B_1 Mangel bei chronischen Athylikern und sein klinisches Korrelat. *Schweiz. Med. Wochenschr.*, **106**, 1466
5. Brubacher, G., Ritzel, G. and Schlettwein-Gsell, D. (1975). Zur Vitaminversorgung ausgewählter Kollektive in Westeuropa. *Lebensmittel und Ernährung*, **28**, 71
6. Brozek, J. (1957). Psychologic effects of thiamine restriction and deprivation in normal young men. *Am. J. Clin. Nutr.*, **5**, 109
7. Deutsche Forschungsgemeinschaft. (1977). *Schwangerschaftsverlauf und Kindesentwicklung*. (Boppard: Boldt)
8. Lemoine, A. (1973). *Contribution à l'étude de la carence vitaminique B_1, B_2, B_6, C à propos d'une Triple Enquête Portant sur 656 Malades Hospitalises.* (Marseille: Thèse)

4
Effects of riboflavin deficiency on erythrocytes

D. I. THURNHAM, F. M. HASSAN and Hilary J. POWERS

In the late 1960s, a method of measuring riboflavin status was introduced[1-3] which was based on the fact that glutathione reductase requires the riboflavin coenzyme, flavin adenine dinucleotide (FAD). Measurements of the activity of erythrocyte glutathione reductase (EGR) *in vitro* with and without FAD are used to calculate an activation coefficient (AC). EGR–AC values less than 1.30 indicate normal riboflavin status while those greater than 1.30 suggest a deficient state.

$$\text{Activation coefficient} = \frac{\text{enzyme activity plus FAD}}{\text{enzyme activity without FAD}}$$

Between 1972 and 1974 we used this test to measure the riboflavin status in approximately 1400 elderly people living in their own homes in two nutritional surveys in the United Kingdom organized by the Department of Health and Social Security. We found that approximately 30% of the subjects were assessed as riboflavin deficient by the EGR test and we were anxious to confirm these data from some of the other information collected, since the surveys also included dietary, clinical, socio-economic and haematological investigations. However, the latter data were not readily available in the early stages due to checking and computerization and we set about its validation in other ways.

For example in the second of the two surveys we obtained information on the breakfast eaten by the subjects on the morning of the venesection. It was possible to show that those subjects who ate vitamin-fortified breakfast cereals (VC) had better riboflavin and, incidentally, thiamin status than those not taking these cereals for breakfast (Table 1)[5]. This should not be interpreted to mean that the dietary intake is having a rapid effect on the EGR test but rather that breakfast habits may be fairly regular and that the subjects identified as eating the breakfast cereals probably did so fairly regularly and had a higher riboflavin intake and better status because of this.

Table 1 Riboflavin status and consumption of vitamin-fortified breakfast cereals (VC) in elderly subjects. An EGR–AC$\geqslant$1.30 is an indication of riboflavin deficiency and percentages are shown in parentheses. (Breakfast means any solid or liquid food taken the morning of venesection by the subjects, who were all over 65 years)

	No. subjects	Breakfast no VC	No breakfast	Breakfast and VC
Total	747	30	87	
With EGR–AC$\gg$1.30	198 (27)	7 (23)	8 (9*)	

*Significantly less than in the other two groups $p < 0.05$

To some extent these findings confirmed that our biochemical data were probably in agreement with the dietary information. However, the large number of subjects who were biochemically deficient in riboflavin by our measurements made us wonder whether there were any differences in the red cells of old people compared to younger subjects which might partly account for our results. For example the activity of glutathione reductase falls as red cells age[6][7] so if there were slightly more old red cells present in elderly people than in younger subjects this might affect the calculation of the activation coefficient.

Table 2 Possible effect of varying erythrocyte glutathione reductase activity on the calculation of the activation coefficient. The difference between the enzyme activities in each case is two units as it is assumed that a particular riboflavin availability in the diet would produce the same degree of unsaturation in the enzyme. The difference between the figures shown for the 'young' and 'old' examples has been exaggerated to illustrate a point (see text)

	Young	Old
Enzyme activity plus FAD	13	8
Enzyme activity no FAD	11	6
Activation coefficient	1.18	1.33

A hypothetical situation is shown in Table 2 where the difference between stimulated and basic activities is the same, that is the apoenzyme to which no FAD is bound, but the mean basic activity in one subject is lower than that in the other. The net effect of this situation is to produce two ACs, one of which is greater than 1.30.

Red cells increase in density as they age[8] and experiments were done therefore to fractionate red cells by density using a discontinuous gradient of Ficoll/Triosil mixtures[9]. Using this technique we were able to separate red cells into nine discrete fractions. Numbers of red cells in each fraction were assessed by measuring the haemoglobin concentration since red cell haemoglobin does not change with age[8].

When we compared the distribution of red cells separated by this technique from persons over 65 with those within the age range 20-50 years, there was no difference[7] (Table 3). However, there were differences when the red cell distribution from riboflavin-deficient subjects was compared with that of normal subjects (Table 4). The number of young red cells in riboflavin-deficient subjects was significantly greater than that in normal subjects. However, this was not associated with a demonstrable reticulocytosis.

Table 3   Haemoglobin in red cell fractions of different densities from young and old persons with normal riboflavin status.

Number of subjects are shown in parentheses. Young subjects were 20-50 years and elderly were over 65 years. Red cells were fractionated using a Ficoll/Triosil discontinuous gradient[9], measured as haemoglobin[31] and content of each fraction expressed as a percentage of the total. Riboflavin status[7] (EGR-AC) was measured on unfractioned blood and all were <1.30

Mean red cell density (g/ml)	Haemoglobin (%) mean ± standard deviation	
	Young (13)	Elderly (7)
1.103	3.3 ± 3.2	4.3 ± 4.2
1.110	14.9 ± 14.6	12.6 ± 12.1
1.116	30.5 ± 30.4	27.4 ± 27.1
1.121	16.5 ± 16.4	14.9 ± 14.7
1.124	10.7 ± 10.6	10.7 ± 10.5
1.127	8.4 ± 8.3	10.9 ± 10.9
1.132	7.5 ± 7.4	11.6 ± 11.2
1.1375	3.1 ± 3.0	6.9 ± 6.7
1.145	4.0 ± 3.9	7.7 ± 7.4

Table 4   Haemoglobin in red cell fractions from subjects with normal and deficient riboflavin status. Details are shown in Table 3. Riboflavin status[7] (EGR-AC) was measured on unfractionated blood

Mean red cell density (g/ml)	Haemoglobin (%) mean ± standard deviation	
	Normal (20) EGR-AC<1.30	Riboflavin deficient (8) EGR-AC≥1.30
1.103	3.7 ± 3.0	10.2 ± 10.8*
1.110	10.2 ± 10.8	24.6 ± 12.7*
1.116	29.4 ± 10.0	31.2 ± 13.6
1.121	15.2 ± 6.7	13.3 ± 3.0
1.124	10.2 ± 6.3	5.7 ± 3.8
1.127	9.3 ± 4.1	7.4 ± 5.1
1.1375	3.7 ± 3.0	1.6 ± 1.0
1.145	4.9 ± 3.8	4.7 ± 3.9

*Significance of difference between normal and deficient groups $p < 0.05$

To investigate this observation much of our further work was done on rats. We showed that red cells appeared to be more fragile *in vitro* than those from control animals. Figures 1 and 2 show the increasing tendency of rat red cells to haemolyse in the presence of hydrogen peroxide or hypotonic saline as the EGR–AC increases[10].

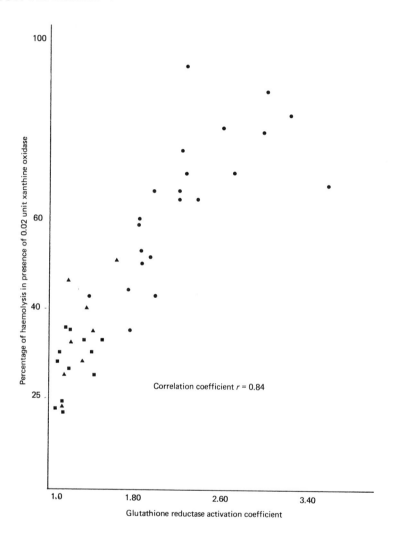

Figure 1 Effect of peroxide on rat erythrocytes in riboflavin deficiency. Red cells from control (■), pair-fed (▲) and riboflavin deficient (●) rats were exposed to a peroxide-generating system[10] in duplicate. Haemoglobin in the supernatant was measured after 45 min and expressed as a percentage of total haemoglobin in the assay. Glutathione reductase activation coefficient represents the riboflavin status[11]

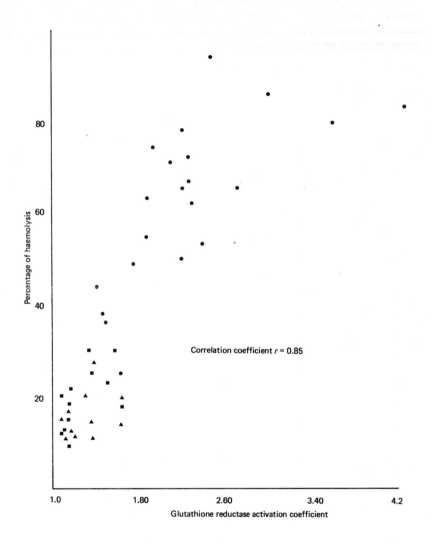

Figure 2 Effect of hypotonic saline on rat erythrocytes in riboflavin deficiency. Details are similar to those in Figure 1 except that erythrocytes were exposed to 0.36% (w/v) sodium chloride[10] for 45 min before measuring haemolysis

Glutathione reductase is necessary for the production of reduced glutathione (GSH) and one of the functions of the latter is as a substrate for glutathione peroxidase. Glutathione peroxidase uses GSH to reduce lipid hydroperoxides in order that the lipid can be removed by the β-oxidation system. Lipid hydroperoxides are formed by the action of oxygen, or oxygen radicals, on double bonds in unsaturated fatty acids and the structural changes produced may affect the permeability of membranes where lipid forms such an essential component[12].

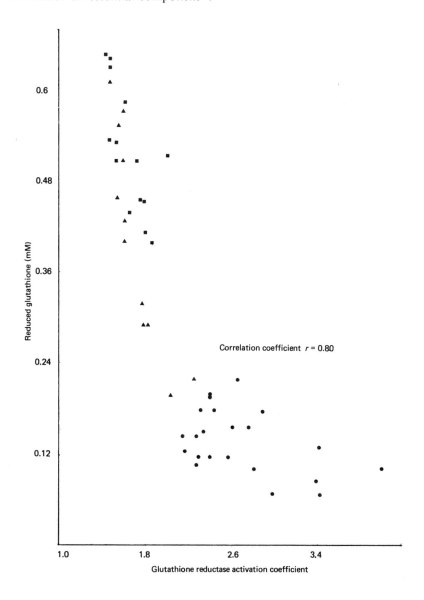

Figure 3 Erythrocyte reduced glutathione (GSH) in riboflavin deficiency in rats. Details are similar to those in Figure 1. Reduced glutathione was measured fluorometrically[32]

RIBOFLAVIN DEFICIENCY AND ERYTHROCYTES

In rats riboflavin deficiency reduces the activity of glutathione reductase and lowers the concentrations of GSH in the red cell. Figure 3 shows that concentrations of GSH are inversely proportional to the EGR–AC[13] and more recently we have shown that there are small but significant increases in both lipid hydroperoxides and malonyldialdehyde in red cells from riboflavin-deficient rats (Table 5). Malonyldialdehyde is an oxidative byproduct of some lipid hydroperoxides and normally its presence cannot be detected[14].

Table 5 Concentrations of lipid hydroperoxides and malonlydialdehyde, and glutathione peroxidase activity in red cells in riboflavin deficiency.

Established methods were used to assay lipid hydroperoxides[24], malonyldialdehyde[25] and peroxidase activity[26]. Results shown are means and standard deviations with numbers of rats shown in parentheses. Results for riboflavin-deficient group are significantly different from both control groups at $p < 0.01$

	Control ad libitum	Control pair-fed	Riboflavin deficient
Total peroxides (μmol/g Hb)	126 ± 36 (9)	116 ± 16 (6)	293 ± 88 (19)
Malonyldialdehyde (μmol/g Hb)	1.9 ± 1.1 (7)	3.0 ± 2.6 (7)	7.9 ± 4.0 (28)
Glutathione peroxidase (IU/g Hb)	4.5 ± 1.0 (9)	4.3 ± 1.2 (6)	7.4 ± 2.1 (27)

One result which was rather surprising at first sight was the *in vitro* activity of erythrocyte glutathione peroxidase. The activity of the enzyme was found apparently to increase in riboflavin deficiency (Table 5). Protein synthesis does not take place in mature red cells and to investigate this enigma the activity of this enzyme was measured in red cells of different ages. Figure 4 shows the expected fall in enzyme activity of erythrocytes from control animals but no fall in enzyme activity occurred in those cells from riboflavin-deficient rats. The fall in enzyme activity which accompanies the physiological ageing of red cells may represent the decay and removal of enzyme which has served its functional life. This would suggest that glutathione peroxidase is not being utilised in red cells from riboflavin-deficient rats, probably because of the fall in GSH concentrations.

A similar situation was found to exist in connection with the control of methaemoglobin concentrations in the red cell. Methaemoglobin is the oxidized form of haemoglobin in which ferrous iron is converted to the ferric state. In the normal state concentrations of red cell methaemoglobin are very low; however, studies in rats have shown that they rise considerably in riboflavin deficiency (Figure 5)[11]. Activity of methaemoglobin reductases is also higher in red cells from the riboflavin-deficient rats though this is only significant ($p < 0.05$) in the case of the NADPH-dependent enzyme (Table 6)

33

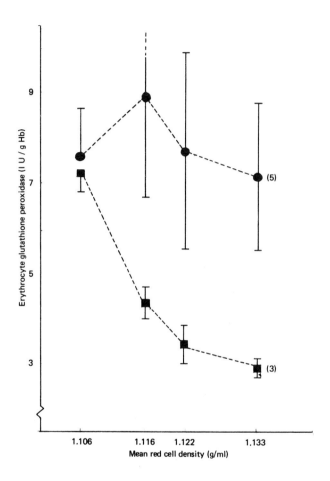

Figure 4 Erythrocyte glutathione peroxidase activity in rats in riboflavin deficiency. Red cells from control (■) and riboflavin-deficient (●) rats were fractionated using a discontinuous gradient[11] and assayed for peroxidase activity[26]. Number of rats is shown in parentheses and vertical lines represent standard deviations

which seems to be behaving in a similar manner to glutathione peroxidase, discussed above (Figure 4).

Table 6 Methaemoglobin reductases in riboflavin deficiency in rats. Published methods were used to assay NADH-methaemoglobin reductase[27] and NADPH-methaemoglobin reductase[28]. Results given are means and standard deviations and numbers of rats are shown in parentheses

	Control ad libitum	Control pair-fed	Riboflavin deficient
NADH- Methaemoglobin reductase (IU/g Hb)	7.4 ± 1.7 (5)	5.6 ± 1.6 (5)	9.5 ± 5.6 (20)
NADPH- Methaemoglobin reductase (IU/g Hb)	3.7 ± 1.5 (8)	3.6 ± 1.5 (9)	6.9 ± 3.3 (22)

The evidence I have presented so far suggests that the metabolism of erythrocytes from riboflavin-deficient rats is less able to withstand the effects of oxidative damage than that of control cells and this may be the explanation for the increased *in vitro* fragility. If this also occurs *in vivo* then erythrocyte lifespan would be shortened. Studies with[51]Cr, however, in riboflavin-deficient rats would suggest that the lifespan of the red cells is not reduced[15]. The ^{51}Cr technique, however, has been criticized[16] as being unable to detect small changes in lifespan and we are at the moment re-investigating the effects of riboflavin deficiency on lifespan using [59]Fe. Unfortunately it is too early to know whether the small changes we are finding will be significant.

We have, however, examined the possibility of there being an effect on lifespan of red cells by fractionation experiments. A series of rats with different degrees of riboflavin deficiency were examined. Table 7 shows that as EGR–AC increased there appeared to be a reduction in the proportion of younger cells and an increase in the proportion of old cells[11]. In agreement with this it has also been shown that there is a decrease in reticulocyte count as the EGR–AC increases and mean packed cell volume, plasma iron and red blood cell count were all lower in the riboflavin-deficient rats although the latter change was not significant (Table 8). On the other hand, iron stores in the liver were raised in riboflavin-deficient rats[13].

There are two possible explanations for the block in erythropoiesis in severe riboflavin deficiency. The enzyme necessary to convert ferric iron in ferritin into the ferrous form to enable it to be transported to the tissues is an oxido-reductase which is flavin-dependent, and it is reported that the activity is reduced[17]. Our results support this finding and suggest that there is impaired mobilization of iron from the liver to haemopoietic tissues. Other workers, however, have not found raised stores of iron in the liver[17 18] but it is not possible to compare the data properly since no measure of riboflavin status was made in these studies. Secondly, plasma thyroxine concentrations are reduced in riboflavin deficiency[19 20] and recent reports have suggested that thyroxine may be necessary for optimal activity of erythropoietin in the stimulation of erythropoiesis[21]. Further work therefore is necessary to

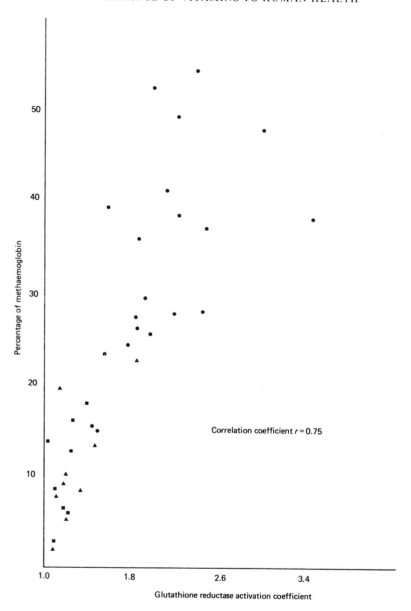

Figure 5 Erythrocyte methaemoglobin in rats in riboflavin deficiency. Details are similar to those in Figure 1 except that methaemoglobin[33] was correlated against riboflavin status

elucidate the mechanism by which anaemia occurs in severe riboflavin deficiency.

Table 7 Influence of riboflavin deficiency on proportions of *cells of different ages in the circulation of rats. Discontinuous density gradient centrifugation[9] was used to fractionate red cells. Data shown are concentration of haemoglobin expressed as percentage of total in red cells fractionated. Control data are given as means and standard deviations; there was no difference between *ad libitum* and pair-fed rats. EGR-AC measurement has been previously described[11]

		Proportions of *cells of densities shown (g/ml)			
	EGR-AC	1.113	1.113–1.119	1.119–1.125	1.125–1.140
Control (9) (*ad libitum*)	1.19±0.07	30.1±12.5	36.1±8.3	20.0±8.2	12.0±5.2
Riboflavin-	1.28	53.2	25.0	15.3	6.5
deficient	1.34	42.0	28.8	12.5	16.7
	1.40	34.5	31.4	25.6	8.6
Individual values	1.44	33.9	28.8	23.8	13.5
	1.49	25.1	40.1	20.7	14.1
	1.53	31.4	40.8	16.2	13.7
	1.60	17.5	33.1	31.5	17.9
	1.64	14.5	28.2	35.7	20.9
	1.67	13.1	27.1	38.2	22.2
	1.78	11.3	32.4	42.4	15.3
	1.80	10.2	27.7	43.0	19.2
	1.80	6.9	32.6	37.6	22.9
	1.86	9.5	24.4	49.4	16.7
	1.88	9.1	30.4	39.0	23.5
	1.90	9.4	25.3	40.3	26.3
	1.95	5.9	16.9	48.1	29.1
	1.97	1.9	16.8	45.5	35.8
	2.13	2.6	28.7	41.2	27.6
	2.47	5.2	22.9	42.8	29.0
	2.61	1.2	18.2	44.8	35.8

Table 8 Haematological measurements in rats with riboflavin deficiency. Published methods were used for assay of plasma iron[29], ferritin iron[30] and non-haem iron[30]. Reticulocytes were stained using brilliant cresyl blue and counted in 1000 cells. A Coulter counter was used for the red cell count and a microhaematocrit centrifuge for the packed cell volume. Results are given as means and standard deviations and numbers of rats are shown in parentheses

Measurement	Control ad libitum	Control pair-fed	Riboflavin deficient	*p
Red cell count ($\times 10^{12}$/litre)	5.53±1.10 (11)	5.36±1.45 (10)	4.84±0.94 (26)	NS
Packed cell volume (%)	47±5 (13)	48±6 (9)	41±5 (43)	<0.001
Reticulocytes (% total)	2.4±0.5 (6)	2.4±0.3 (6)	1.3±0.6 (19)	<0.001
Plasma iron (µg/100 ml)	242±47 (8)	221±51 (5)	191±24 (13)	<0.01
Ferritin iron (µg/g liver)	71±13 (8)	64±15 (8)	121±28 (16)	<0.001
Non-haem iron (µg/g liver)	111±22 (8)	108±19 (8)	192±51 (24)	<0.001

* Significance of difference between *ad libitum* control versus riboflavin-deficient group. NS = not significant

Further examination of Table 7 shows also that some of the rats had EGR–ACs in the control range. When we first obtained these data we believed these animals were marginally deficient and the fact that the proportion of younger cells in these rats was greater than that found in the control animals, was similar to our results in marginally deficient human subjects. However, there was clinical evidence of riboflavin deficiency of varying severities in all these rats and the low EGR–ACs may mean that these rats had refected and were recovering. The increased proportion of young cells in these 'marginally' deficient animals is therefore not necessarily comparable with the data obtained from marginally deficient human subjects but we have not yet fractionated blood from riboflavin-deficient rats before the appearance of clinical signs to confirm this point.

It does, however, appear at the present time that the effect of riboflavin deficiency on the production of erythrocytes depends on the severity of the deficiency. A marginal deficiency probably stimulates erythropoiesis while a severe deficiency blocks it. These observations may explain some of the conflicting published reports concerning riboflavin deficiency and anaemia[22].

In terms of human nutrition both situations are important. Marginal riboflavin deficiency occurs extensively throughout the world[23] and in the marginal situation, in particular, an increase in erythropoiesis will increase the body's requirements for all other haemopoietic factors. Iron, folate, vitamin B_{12} etc. may already be in short supply: thus it is possible that anaemia, which is often as common as riboflavin deficiency in the same communities, may be partly attributable to poor riboflavin status.

To what extent is biochemical riboflavin deficiency in elderly people in this country associated with an increased incidence of anaemia? Approximately 12% of the subjects in the 1972 survey[4] were classed as anaemic and there was an inverse correlation between haemoglobin and the EGR–AC ($r = -0.19$, $p < 0.05$, $n = 195$), but in the women only. However, although the number of subjects with anaemia in the riboflavin-deficient group was increased in both men and women, the differences were not significant. Since the prevalence of anaemia was fairly low, the absence of a difference between those subjects who were riboflavin-deficient and those who were biochemically normal may simply mean that the majority of the riboflavin-deficient elderly subjects had an adequate intake of other haemopoietic factors. Therefore riboflavin deficiency as seen in this country is probably not a problem in the aetiology of anaemia in most cases.

SUMMARY

Studies on elderly subjects in the United Kingdom using the glutathione reductase test and the activation coefficient (AC) as a measure of riboflavin status suggested that approximately 30% of the subjects were riboflavin deficient.

Investigation into whether mean erythrocyte glutathione reductase activity was lower in the blood of the old people and might influence the calculation of the AC provided no evidence to support the suggestion. However, discontinuous density gradient fractionation of red cells from riboflavin-deficient humans (AC $\gg$ 1.30) indicated an increased number of young and fewer old erythrocytes by comparison with biochemically normal subjects.

Studies done on riboflavin-deficient rats have shown that erythrocytes are more fragile *in vitro*, have a lower concentration of reduced glutathione, show evidence of oxidative damage and have a reduced mean lifespan. In spite of this erythropoiesis appears to be reduced in rats with severe riboflavin deficiency, although in the marginal state it is possible that erythropoiesis may be increased.

Riboflavin deficiency occurs extensively in areas of the world where anaemia is also very common. The possibility that even marginal riboflavin may be a contributory factor in the development of anaemia is discussed.

Acknowledgments

D.I.T. acknowledges support from the Department of Health and Social Security and would also like to thank the Medical Research Council and Central Research Fund, London University, for a grant to purchase a Beckman SW 27.1 swinging bucket rotor.

References

1. Bamji, M. S. (1969). Glutathione reductase activity in blood cells and riboflavin nutritional status in humans. *Clin. Chim. Acta*, **26**, 263
2. Beutler, E. (1969). The effect of flavin coenzymes on the activity of erythrocyte enzymes. *Experientia*, **25**, 804
3. Glatzle, D., Körner, W. F., Christeller, S., and Wiss, O. (1970). Methods for the detection of a biochemical riboflavin deficiency. *Int. J. Vit. Nutr. Res.*, **40**, 166
4. Department of Health and Social Security. (1978). *Nutrition and Health in Old Age*. A follow-up study made in 1972/73 of the 1967/68 sample. (London: HMSO)
5. Thurnham, D. I. (1977). The influence of breakfast habits on vitamin status in the elderly. *Proc. Nutr. Soc.*, **36**, 97A
6. Ganzoni, A. M., Barras, J. P. and Marti, H. R. (1976). Red cell ageing and death. *Vox Sang.*, **30**, 161
7. Powers, H. J. and Thurnham, D. I. (1976). Influence of red cell age on the measurement of riboflavin status. *Nutr. Metab.*, **21** (Suppl. 1), 155
8. Allison, A. C. and Burn, G. P. (1955). Enzyme activity as a function of age in the human erythrocyte. *Br. J. Haematol.*, **1**, 291
9. Turner, B. M., Fisher, R. A. and Harris, M. (1974). The age-related loss of activity of four enzymes in the human erythrocyte. *Clin. Chim. Acta*, **50**, 85
10. Hassan, F. M. and Thurnham, D. I. (1977). Influence of riboflavin status on red blood cell fragility in rats. *Proc. Nutr. Soc.*, **36**, 64A
11. Hassan, F. M. and Thurnham, D. I. (1977). Effect of riboflavin deficiency on the metabolism of the red blood cell. *Int. J. Vit. Nutr. Res.*, **47**, 349
12. Mengel, C. H. and Kann, H. E., Jr. (1966). Effects of *in vivo* hyperoxia on erythrocytes. III. *In vivo* peroxidation of erythrocyte lipid. *J. Clin. Invest.*, **45**, 1150
13. Hassan, F. M. (1978). Erythrocyte pathophysiology of riboflavin deficiency in rats. PhD thesis: University of London

14. Schauenstein, E., Esterbauer, H. and Zollner, H. (1977). *Aldehydes in Biological Systems*, p. 133. (London: Pion Ltd.)
15. Beutler, E. and Srivastava. (1970). Relationship between glutathione reductase activity and drug induced haemolytic anaemia. *Nature*, **226**, 755
16. Belcher, E. H. and Harris, E. B. (1959). Studies of red cell life span in the rat. *J. Physiol.*, **146**, 217
17. Sirivich, S., Driskell, J. and Frieden, E. (1977). NADH–FMN oxidoreductase activity and iron content of organs from riboflavin and iron deficient rats. *J. Nutr.*, **107**, 739
18. Zaman, Z. and Verwilghen, R. L. (1977). Effect of riboflavin deficiency on activity of NADH–FMN-oxido-reductase (ferriductase) and iron content of rat livers. *Biochem. Soc. Trans.*, **5**, 306
19. Nolte, J., Brdiczka, D. and Staudte, H. W. (1972). Effect of riboflavin deficiency on metabolism of the rat in hypothyroid and euthyroid state. *Biochim. Biophys. Acta*, **268**, 611
20. Rivlin, R. S., Mendez, C. and Langdon, R. G. (1968). Biochemical similarities between hypothyroidism and riboflavin deficiency. *Endocrinology*, **83**, 461
21. Golde, D. W., Bersch, N., Chopra, I. J. and Cline, M. J. (1977). Thyroid hormones stimulate erythropoiesis *in vitro*. *Br. J. Haematol.*, **37**, 173
22. Lane, M., Smith, F. E. and Alfrey, C. P. Jr. (1975). Experimental dietary and antagonist-induced human riboflavin deficiency. In: R. S. Rivlin (ed.). *Riboflavin*, p. 269 (London: Plenum Press)
23. W.H.O. (1965). Requirements of vitamin A, thiamine, riboflavin and niacin. *Techn. Rep. Ser.*, no. 362
24. Heath, R. L. and Tappel, A. L. (1976). A new sensitive assay for the measurement of hydroperoxide. *Anal. Biochem.*, **76**, 184
25. Mengel, C. E., Kann, H. E., Jr. and Merriwether, W. D. (1967). Studies of paroxysmal nocturnal hemoglobinuria erythrocytes: Increased lysis and lipid peroxide formation by hydrogen peroxide. *J. Clin. Invest.*, **46**, 1715
26. Beutler, E. (1971). *Red Cell Metabolism. A Manual of Biochemical Methods.* (New York: Grune and Stratton)
27. Hegesh, E., Calmanovici, N. and Avron, M. (1956). New method for determining ferri-haemoglobin reductase (NADH-methaemoglobin reductase) in erythrocytes. *J. Lab. Clin. Med.*, **44**, 668
28. Huennekens, F. M., Caffrey, R. W., Basford, R. E. and Gabria, B. W. (1957). Erythrocyte metabolism. IV. Isolation and properties of methaemoglobin reductase. *J. Biol. Chem.*, **227**, 261
29. Bothwell, T. M. *et al.* (1971). Proposed recommendation for measurement of serum iron in human blood. International Committee for Standardization in Hematology. *Br. J. Haematol.*, **20**, 451
30. Foy, A. L., William, H. L., Cortell, S. and Conrad, M. E. (1967). A modified procedure for the determination of non-haem iron in tissue. *Anal. Biochem.*, **18**, 559
31. Wootton, I. D. P. (1964). *Microanalysis in Medical Biochemistry.* 4th ed. p. 119. (London: Churchill Livingston)
32. Hissin, P. J. and Hilf, R. (1976). A fluorometric method for determination of oxidised and reduced glutathione in tissues. *Anal. Biochem.*, **74**, 214
33. Evelyn, K. A. and Malloy, H. T. (1938). Microdetermination of oxyhaemoglobin, methaemoglobin and sulfhaemoglobin in a single sample of blood. *J. Biol. Chem.*, **126**, 655

5
Nutritional and biochemical aspects of vitamin B_{12}

A. V. HOFFBRAND

THE COBALAMINS

The term 'vitamin B_{12}' refers to a chemically defined compound, cyanocobalamin, but it is also used to describe a small group of closely related compounds, the cobalamins, with the same basic structure, consisting of a nucleotide containing a base 5, 6-dimethyl-benziminazole attached to a sugar,

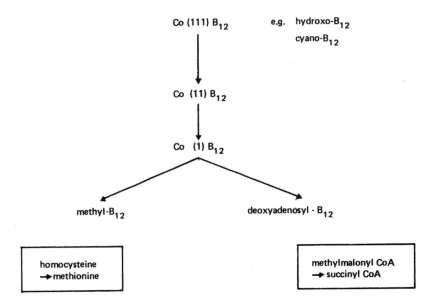

Figure 1 Pathway of reduction of hydroxo- or cyano- cobalamin to the two active coenzyme vitamin B_{12} derivatives, methyl- and deoxyadenosyl-B_{12}. Reactions in which the coenzymes are involved are shown in a box below each coenzyme.

ribose-3-phosphate, set at right angles to a corrin ring with a cobalt atom at its centre. The vitamin was first crystallized in 1948 as cyanocobalamin in which a –CN group is attached to the cobalt atom but this form only occurs in trace amounts in natural materials. The main forms in nature and indeed in human tissues are now known to be methyl-, deoxyadenosyl- and hydroxo-cobalamins (methyl-, ado and hydroxo-B_{12}) (Figure 1). Cyano- and hydroxo-cobalamins are stable oxidized compounds with the cobalt atom in the Co (III) state, so-called cob(III)alamins, whereas the methyl- and deoxy-adenosyl compounds are reduced, cob(I)alamin derivatives which are easily oxidized by light to the hydroxo-compound. Methyl-B_{12} is involved in conjunction with N5-methyltetrahydrofolate in the methylation of homo-cysteine to methionine, whereas deoxydenosyl-B_{12} is concerned in the isomerization of methylmalonyl CoA (a metabolite in propionate metabolism) to succinyl CoA. Hydroxocobalamin is used in the therapy of vitamin B_{12} deficiency while cyanocobalamin is used as its radioactive [57]Co or [58]Co derivatives for testing vitamin B_{12} absorption or for *in vitro* tests requiring radioactive vitamin B_{12}.

DIETARY SOURCES

The vitamin is synthesized in nature solely by micro-organisms. Plants do not contain any detectable vitamin B_{12} whereas all animal tissues do. Animals gain the vitamin by eating other animal tissues, by absorption of the vitamin synthesized in the intestine by micro-organisms, or by eating vegetables, nuts and fruits which have been contaminated by micro-organisms.

Humans gain the vitamin by eating animal foods, the highest content being in liver (approximately 100 $\mu g/g$) and kidney, but fish, muscle meat, poultry, cheese, eggs, butter and milk all contain some vitamin B_{12}, in milk approximately 6 $\mu g/l$. Minimal adult daily losses and thus requirements have been variously estimated at 0.5–3.0 μg, 1–2 μg, generally considered to be usual. A normal mixed daily diet contains 5–30 μg, the exact amount depending largely on the amount of animal protein (and thus cost) of the diet. Cooking leads to little if any loss of the vitamin, unless alkaline conditions are used.

VITAMIN B_{12} ABSORPTION

Absorption of vitamin B_{12} depends on its release from protein binding in food, attachment to a glycoprotein, intrinsic factor (IF), secreted by the parietal cells of the stomach, attachment of the complex to receptors on the ileal cell brush border surface, and transfer (after several hours) of the vitamin to portal plasma attached (probably in the ileal cell) to a polypetide transport protein, transcobalamin II[1]. The absorption of vitamin B_{12} is normally limited by the number of ileal IF receptors, vast excesses of IF being secreted daily. At a dose of 1 μg of crystalline vitamin B_{12}, 50–70% is absorbed but with larger doses proportionately less is absorbed. A very small proportion (less than 1%) of vitamin B$_{12}$ may be absorbed through the buccal, gastric and upper intestinal mucosae by a non-intrinsic factor mechanism. This inefficient absorption is passive and rapid whereas absorption of vitamin B$_{12}$ by the IF mechanism is active, requires calcium ions, a neutral pH, and is slow. Gastric juice also contains a non-IF vitamin B$_{12}$ binding protein, a so called 'R' protein, closely related to similar glycoproteins in other body fluids (plasma, bile, milk, saliva, tears, etc.). Vitamin B$_{12}$ bound to this protein must be freed before it is available for absorption. Pancreatic trypsin appears to be necessary for this and so patients with chronic pancreatitis may absorb vitamin B$_{12}$ poorly[2]. Other subjects with achlorhydria, e.g. due to cimetidine therapy or atrophic gastritis, may also fail to absorb protein-bound vitamin B$_{12}$ even though their absorption of crystalline vitamin B$_{12}$ is unimpaired[3]. The role of the gastric R protein may be to impede absorption of pseudo-vitamin B$_{12}$ compounds that occur in nature[4].

An enterohepatic circulation for vitamin B$_{12}$ has been estimated to contain up to 6 μg of vitamin B$_{12}$ daily. This vitamin B$_{12}$ together with vitamin released from sloughing intestinal cells is presumably reabsorbed after attachment to IF. Loss of all this vitamin B$_{12}$ because of malabsorption may explain the greater clearance of the vitamin in untreated pernicious anaemia than in normal subjects or vegans[5].

VITAMIN B$_{12}$ DEFICIENCY

Severe deficiency causes megaloblastic anaemia, by interfering with folate metabolism, and it may also cause a neuropathy and glossitis and other symptoms due to epithelial cell damage. Sterility may occur in either sex. Widespread reversible melanin pigmentation occurs in a small proportion of patients. Inadequate dietary intake arises mainly in vegans and by far the largest group are Hindus. Many more show mild abnormalities, e.g. low serum vitamin B$_{12}$ levels, than develop overt megaloblastic anaemia[10]. A very poor diet, lacking animal protein, may also rarely contribute to megaloblastic anaemia in non-vegans. In the case of dietary deficiency, the anaemia will respond to 1–2 μg vitamin B$_{12}$ by mouth[11]. More frequently in Western countries, vitamin B$_{12}$ deficiency is due to malabsorption of the vitamin either because of lack of intrinsic factor (as in adult pernicious anaemia) or because of a small intestinal defect, e.g. the intestinal stagnant-loop syndrome or after an ileal resection. In these cases parenteral vitamin B$_{12}$ therapy is needed to elicit an adequate response.

INACTIVATION OF VITAMIN B$_{12}$

There is no syndrome of increased vitamin B$_{12}$ consumption. Even in pregnancy, the demands for the vitamin for the fetus, 60–70 μg, are small compared with adult body stores of 2–3 mg, and the body is unable to degrade

Table 1   Causes of megaloblastic anaemia due to vitamin B_{12} deficiency (A) or disturbances of vitamin B_{12} metabolism (B)

(A)	*Causes of vitamin B_{12} deficiency (depletion of body stores of vitamin B_{12})*			
(i)	Inadequate intake — vegans			
(ii)	Malabsorption	— (a)	gastric causes (deficiency of intrinsic factor)	e.g. pernicious anaemia, (partial) gastrectomy
		— (b)	intestinal causes	e.g. intestinal stagnant-loop, tropical sprue, ileal resection, specific malabsorption with proteinuria etc.

N.B.   Many causes of malabsorption of vitamin B_{12}, e.g. chronic pancreatitis, are not sufficiently severe or prolonged to lead to serious vitamin B_{12} deficiency

(B)	*Disturbances of vitamin B_{12} metabolism*	
(i)	*Congenital*	— failure of formation of ado-B_{12} and/or methyl-B_{12} — transcobalamin II deficiency
(ii)	*Acquired*	— nitrous oxide inhalation — ?? inactivation as cyano-B_{12}

the vitamin. Thus, it takes some 3–4 years to totally deplete the body of vitamin B_{12} from all the causes of B_{12} deficiency – inadequate dietary intake or malabsorption. However, it has recently been shown that inhalation of nitrous oxide may inactivate vitamin B_{12}, particularly the methyl-B_{12} derivative and cause megaloblastic change with normal serum levels of vitamin B_{12} and folate, although responsive *in vitro* and *in vivo* to large doses of vitamin B_{12}[6]. Prolonged exposure in the experimental animal and possibly in humans may lead to a neuropathy similar to that due to severe depletion of vitamin B_{12} due to malabsorption of the vitamin[7 8].

The theory that body vitamin B_{12} could be inactivated in the cyano form due to excess dietary intake of cyanate, e.g. in cassava, or to heavy smoking, has not been substantiated by direct measurements of plasma and tissue vitamin B_{12} compounds.

Two rare congenital vitamin B_{12} responsive, non B_{12}-deficient syndromes have been described. In one, the infant is born lacking transcobalamin II[9]. Megaloblastic anaemia develops within a few weeks of birth due to failure of plasma B_{12} to enter cells; large parenteral doses of vitamin B_{12}, which raise plasma levels extremely high, correct the anaemia, and sufficient vitamin B_{12} enters the cells by passive diffusion to correct the intracellular deficit. In the other congenital syndrome, failure to synthesize ado-B_{12} leads to methylmalonic aciduria, responsive to large doses of vitamin B_{12}.

DIAGNOSIS OF VITAMIN B_{12} DEFICIENCY

This is suspected from the clinical history and examination, and from the blood and bone marrow appearances. Measurement of the serum vitamin B_{12} level is the most usual method of diagnosing vitamin B_{12} deficiency in the laboratory. This can be performed microbiologically or by isotope assay but recent studies have suggested that isotope assays, as currently performed with non-IF binding proteins, detect analogues of vitamin B_{12} in plasma which are neither microbiologically nor haematologically active. Thus the isotope assays give falsely high values, and may fail to detect vitamin B_{12} deficiency, unless, it is suggested, pure IF is used as the binding protein or the non-IF binders are first saturated with these vitamin B_{12} analogues[12 13].

Measurement of methylmalonic acid is a less convenient and less sensitive guide to vitamin B_{12} deficiency. Vitamin B_{12} absorption tests are important in elucidating the cause of vitamin B_{12} deficiency, whether dietary, due to lack of IF or due to an intestinal lesion (when absorption cannot be corrected by additional IF). There is some evidence that the test as carried out with labelled crystalline cyano-B_{12} fails to detect malabsorption of vitamin B_{12} in those subjects, e.g. following partial gastrectomy who are unable to release vitamin B_{12} in food from protein[14]. Tests using protein-bound ^{57}Co-B_{12} have therefore been devised, but are not widely used since for most patients the tests with the crystalline compound give good discrimination between normal and patients with deficiency due to malabsorption of the vitamin[15].

References

1. Chanarin, I., Muir, M., Hughes, A. and Hoffbrand, A. V. (1978). Evidence for intestinal origin of transcobalamin II during vitamin B_{12} absorption. *Br. Med. J.*, **1**, 1453
2. Allen, R. H., Seetharam, B., Podell, E. and Alpers, D. H. (1978). Effect of proteolytic enzymes on binding of cobalamin to R protein and intrinsic factor: *in vitro* evidence that a failure to partially degrade R protein is responsible for cobalamin malabsorption in pancreatic insufficiency. *J. Clin. Invest.*, **61**, 47
3. King, C., Leibach, J. and Toskes, P. (1977). Food-bound vitamin B_{12} malabsorption — a clinically important cause of vitamin B_{12} deficiency. (Abstract). *Gastroenterology*, **72**, 1080
4. Kolhouse, J. F. and Allen, R. H. (1977). Absorption, plasma transport, and cellular retention of cobalamin analogues in the rabbit: Evidence for the existence of multiple mechanisms that prevent the absorption and tissue dissemination of naturally occurring cobalamin analogues. *J. Clin. Invest.*, **60**, 1381
5. Amin, A., Spinks, T., Ranecar, A., Short, M. D. and Hoffbrand, A. V. (1978). Long-term clearance of ^{57}Co-cyanocobalamin in vegans and pernicious anaemia. *Clin. Sci. Mol. Med.* (In press)
6. Amess, J. A. L., Burman, J. F., Rees, G. M., Nancekievill, D. G. and Mollin, D. L. (1978). Megaloblastic haemopoiesis in patients receiving nitrous oxide. *Lancet*, **ii**, 339
7. Dinn, J. J., McCann, S., Wilson, P., Reed, B., Weir, D. and Scott, J. (1978). Animal model for subacute combined degeneration. *Lancet*, **ii**, 1154
8. Layzer, R. B. (1978). Myeloneuropathy after prolonged exposure to nitrous oxide. *Lancet*, **ii**, 1227
9. Hakami, N., Neiman, P. E., Canellos, G. P. and Lazerson, J. (1971). Neonatal megaloblastic anaemia due to inherited transcobalamin II deficiency in two siblings. *N. Engl. J. Med.*, **285**, 1163

10. Roberts, P. D., James, H., Petrie, A., Morgan, J. O. and Hoffbrand, A. V. (1973). Vitamin B_{12} status in pregnancy among immigrants to Britain. *Br. Med. J.*, **3**, 67
11. Stewart, J. S., Roberts, P. D. and Hoffbrand, A. V. (1970). Response of dietary vitamin B_{12} deficiency to physiological oral doses of cyanocobalamin. *Lancet*, **ii**, 542
12. Kolhouse, J. F., Kondo, H., Allen, N. C., Podell, E. and Allen, R. H. (1978). Cobalamin analogues are present in human plasma and can mask cobalamin deficiency because current radioisotope dilution assays are not specific for true cobalamin. *N. Engl. J. Med.*, **299**, 785
13. Cooper, B. A. and Whitehead, V. M. (1978). Evidence that some patients with pernicious anaemia are not recognised by radio dilution assay for cobalamin in serum. *N. Engl. J. Med.*, **299**, 816
14. Doscherholmen, A., McMahon, J. and Ripley, D. (1976). Inhibitory effect of eggs on vitamin B_{12} absorption: description of a simple ovalbumin ^{57}Co-vitamin B_{12} absorption test. *Br. J. Haematol.*, **33**, 261
15. Mollin, D. L. and Waters, A. H. (1968). The study of vitamin B_{12} absorption using labelled cobalamins. *Medical Monographs*, **6**, The Radiochemical Centre, Amersham, England

6
Folic acid

I. CHANARIN

Folic acid is an essential nutrient that is widely distributed in foods of both plant and animal origin.

CHEMISTRY

Folate polyglutamate is the predominant analogue in cells and 5-methyl-tetrahydrofolate is the extracellular and transport form present in plasma and cerebrospinal fluid (CSF). In man the pentaglutamate, one with five glutamic acid residues, predominates. In addition natural folates are reduced to the tetrahydro- or dihydrofolate level and most carry an additional single carbon either as a formyl or methyl group[1].

DISTRIBUTION

Folate is widely distributed in foodstuffs and most foods contribute some folate. Cooking in large volumes of fluid, however, leaches the soluble folates into the fluid and this has been used as a means of producing a folate free diet[2]. Not only does the folate pass into the liquid phase but much of it is damaged by oxidation, a point illustrated by Hurdle in studying folate in cabbage, when the bulk was destroyed by heat[3]. On the other hand, in the presence of reducing conditions, as may be the case in fresh food or when reducing agents such as ascorbate are added, the folate is stable and will resist boiling. There is a close parallelism between the presence of ascorbate and folate in food. Both are labile and susceptible to oxidation. A diet low in ascorbate is also likely to be low in folate and clinical scurvy is almost invariably accompanied by folate deficiency, often of sufficient severity to produce an accompanying mega-loblastic anaemia.

UPTAKE AND ABSORPTION

A mixed Western type diet contains 150–300 μg folate daily[4]. In 1951 the US Department of Agriculture produced a booklet on the folic acid content of foods[5]. These values are now known to be between 2 and 20 times too low because preparations for the assays were made in the absence of ascorbate[6]. More reliable food tables were produced by Hoppner and his colleagues in Canada[7]. There is still some uncertainty about the availability for absorption of dietary folate[1]. Natural folate monoglutamates have been shown to be absorbed virtually completely using faecal excretion methods with tritium labelled folate. About half the folate in a diet is monoglutamate, largely the result of breakdown of polyglutamates by enzymes that remove the glutamic acid chain. At one stage it was thought that polyglutamates were very poorly absorbed. They are absorbed as monoglutamates and during absorption the glutamic acid residues in excess of one are removed. The enzyme bringing this about has been called a conjugase and is a lysosomal enzyme active at pH 4.5[8]. If this is the site of cleavage then the relatively large polyglutamate molecule must get itself into the cell and into the lysosome where these acid hydrolytic enzymes function. On the other hand Halsted and his group have claimed that there is a brush border conjugase active at more neutral pH[9]. This awaits confirmation. Be that as it may, limited data show that polyglutamate is less well absorbed than equimolar amounts of monoglutamate. Absorption ranges from 30 to 70% of the total polyglutamate but reliable quantitative data are not available[1].

The mechanism by which natural folates are absorbed ensures that only 5-methyltetrahydrofolate is delivered to portal blood[10]. Unphysiological folate – *viz.* pteroylmonoglutamic acid – is absorbed largely unchanged in the gut and further modifications to the molecule therefore take place in tissues such as the liver.

MODE OF ACTION

Folate is concerned with a variety of synthetic reactions in purine and pyrimidine synthesis. Because these compounds form the bases in DNA and RNA the high folate requirement in growing tissues is evident. Interference of DNA synthesis in malignant cells forms the basis of the folate antagonist action of methotrexate.

FOLATE DEFICIENCY

Dietary deficiency

Inadequate intake of poor quality, overcooked food leads to nutritional folate deficiency. This is widespread in developing countries and in more affluent societies it affects the elderly[1]. It may be assessed by noting the frequency of megaloblastic anaemia due to folate deficiency and by measuring serum and red cell folate levels in population groups.

In man a relatively folate-free diet rapidly produces a negative folate balance so that the serum folate level, maintained in part by folate absorbed from the gut, falls to low levels within the first 2 weeks[11]. There is a slower fall in red cell folate as dilution of old red cells by new ones of low folate content occurs and it only becomes abnormally low in a healthy adult after 17 weeks. At about this time, too, morphological changes may appear in marrow and blood. These changes occur much sooner in those already folate deficient, for example alcoholics[12]. In the UK the frequency of low red cell folates is variable, but was found to be 8% in both Sheffield and London[1]. No evidence of deficiency was found in Edinburgh.

Megaloblastic anaemia has occurred in patients being maintained on parenteral nutrition over several weeks[13] as well as in children receiving artificial diets for treatment of phenylketonuria.

Infancy

There is an increased frequency of folate deficiency among premature infants. Gray and Butler[14] described megaloblastic anaemia in three infants whose birthweights were 850, 1000 and 1150 g respectively. All responded to folate. Strelling et al.[15] found megaloblastic anaemia in at least 7 and possibly 11 out of 54 premature infants.

The decline in serum and red cell folate is greater in premature than in full term neonates. This is probably due to reduced folate stores accumulated by the premature infant in utero and, in the past, to an inadequate folate content of reheated milk feeds. Pooled human breast milk used to feed premature infants had only 3 μg folate/l instead of a normal value of about 50 μg/l[16]. There is also a relatively large folate requirement due to a high growth rate.

Rare examples of megaloblastic anaemia in infants reared on goat's milk have been reported[1]. The folate content of goat's milk is only 10% of that of human or cow's milk. Infants should not be given goat's milk as the major item of diet without folate supplementation.

Pregnancy

Pregnancy is the other situation in which there is an increased requirement for folate and if it is not met from dietary sources folate deficiency and megaloblastic anaemia ensues. In many countries the regular administration of folate supplements during pregnancy has largely eliminated folate deficiency and megaloblastic anaemia.

Before the widespread use of prophylactic folate one quarter of marrow samples taken in pregnancy showed megaloblastic changes (Table 1) and in two studies in South India this exceeded half of all the samples. The megaloblastic changes were eliminated by folate supplementation and in relatively few patients were similar changes noted on peripheral blood examination

THE IMPORTANCE OF VITAMINS TO HUMAN HEALTH

where iron deficiency was the usual finding.

In order to prevent a fall in the serum or red cell folate level in pregnancy an average of 100 μg folate daily has to be added to the normal folate intake. To meet the needs of all pregnant women a supplement of 200–300 μg folate daily is adequate.

Table 1 Percentage frequency of megaloblastic marrow changes in pregnancy[1<

UK	25
Ireland	30
Canada	25
USA (Texas)	24
Nigeria	30
South Africa (negro population)	25
South India	54 and 60

Alcohol

Finally folate deficiency is commonly associated with alcoholism. The evidence is that this is due to associated dietary deficiency of folate and not to any special effect of alcohol. It affects spirit drinkers much more than beer drinkers since beer contains some 100 μg folate/l[17].

References

1. Chanarin, I. (1979). *The Megaloblastic Anaemias*, Second Edition, (Oxford: Blackwell)
2. Herbert, V. (1963). A palatable diet for producing experimental folate deficiency in man. *Am. J. Clin. Nutr.*, **12**, 17
3. Hurdle, A. D. F. (1967). The folate content of a hospital diet. MD thesis: University of London.
4. Moscovitch, L. F. and Cooper, B. A. (1973). Folate content of diets in pregnancy: comparison of diets collected at home and diets prepared from dietary records. *Am. J. Clin. Nutr.*, **26**, 707
5. Toepfer, E. W., Zook, E. G., Orr, M. L. and Richardson, L. R. (1951). Folic acid contents of foods. Microbiological assay by standardization methods and compilation of data from the literature. *Agriculture Handbook No. 29*, (US Department of Agriculture)
6. Thenen, S. W. (1975). Food folate values. *Am. J. Clin. Nutr.*, **28**, 1341
7. Hoppner, K., Lampi, B. and Perrin, D. E. (1972). The free and total folate activity in foods available on the Canadian market. *J. Inst. Can. Sci. Technol. Aliment.*, **5**, 60
8. Hoffbrand, A. V. and Peters, T. J. (1969). The subcellular localization of pteroylpoly-glutamate hydrolase and folate in guinea pig intestinal mucosa. *Biochim. Biophys. Acta*, **192**, 479
9. Reisenauer, A. M., Krumdieck, C. L. and Halsted, C. H. (1977). Folate conjugase: Two separate activities in human jejunum. *Science*, **198**, 196
10. Perry, J. and Chanarin, I. (1970). Intestinal absorption of reduced folate compounds in man. *Br. J. Haematol.*, **18**, 329
11. Herbert, V. (1962). Experimental nutritional folate deficiency in man. *Trans. Assoc. Am. Physicians.*, **75**, 307
12. Eichner, E. R. and Hillman, R. S. (1971). The evolution of anaemia in alcoholic patients. *Am. J. Med.*, **50**, 218

13. Beard, M. E. J., Hatipov, C. S. and Hamer, J. W. (1978). Acute marrow folate deficiency during intensive care. *Br. Med. J.*, **1**, 624
14. Gray. O. P. and Butler, E. B. (1965). Megaloblastic anaemia in premature infants. *Arch, Dis. Child.*, **40**, 53
15. Strelling, M. K., Blackledge, G. D., Goodall, H. B. and Walker, C. H. M. (1966). Megaloblastic anaemia and whole-blood folate levels in premature infants. *Lancet*, **i**, 898
16. Roberts, P. M., Arrowsmith, D. E., Rau, S. M. and Monk-Jones, M. E. (1969). Folate state of premature infants. *Arch, Dis. Child.*, **44**, 637
17. Wu, A., Chanarin, I., Slavin, G. and Levi, A. J. (1975). Folate deficiency in the alcoholic — its relationship to clinical and haematological abnormalities, liver disease and folate stores. *Br. J. Haematol.*, **29**, 469

7
Clinical biochemistry of vitamin B$_6$

S. B. ROSALKI

Vitamin B$_6$ is unique in the number and diversity of the metabolic reactions in which it is involved and the number of diseases in which its deficiency has been reported. Its clinical biochemistry is here reviewed.

B$_6$ METABOLISM

Vitamin B$_6$ or pyridoxine is the collective name given to three substituted pyridines: an alcohol, pyridoxol, formerly known as pyridoxine and found especially in plants, an aldehyde, pyridoxal, and an amine, pyridoxamine. The aldehyde and amine are the principal forms found in animal tissue.

Dietary B$_6$ is absorbed from the intestine and passes to the liver where the different vitamin forms are interconverted and phosphorylated to yield pyridoxal phosphate. Pyridoxal phosphate circulates in the plasma bound to albumin and passes to the tissue stores. Removal from albumin and dephosphorylation are required before transport into the cell, and rephosphorylation occurs after cell entry. The main end-product of vitamin B$_6$ metabolism is 4-pyridoxic acid, which derives largely from the aldehyde form and is excreted in the urine[1][2].

The major metabolic function of vitamin B$_6$ is as an enzyme component or coenzyme. The principal active form of the vitamin is pyridoxal phosphate, which acts as a coenzyme in more than 60 enzyme reactions, including those decarboxylation, deamination, transamination, transmethylation and transulphuration reactions which control amino acid synthesis and transport and the production of bioamines such as adrenaline, serotonin and other neurotransmitters. Other coenzyme functions of the vitamin control the formation of the vitamin nicotinic acid from the amino acid tryptophan, the breakdown of glycogen, fatty acid synthesis, and the formation of porphyrin and the

53

haem portion of haemoglobin from amino acid precursors[3][4]. Worth noting is the fact that vitamin B_6 and other vitamins are interrelated. Riboflavin acts as a coenzyme for pyridoxal phosphate synthesis, which in turn controls nicotinic acid production.

DIETARY B_6

Vitamin B_6 is a water soluble vitamin and is found principally in wheat, corn, meat and fish, and to a lesser extent in milk, eggs, cheese and green vegetables. The daily requirement for the vitamin varies with protein intake and approximates to 20–30 $\mu g/g$ protein, representing an average requirement of some 2 mg per day in the adult[5][6]. There is an increased requirement for the vitamin to support the growth needs of childhood, pregnancy and lactation. Some synthesis of the vitamin takes place in the gut bacteria but this is generally believed to be non-available for absorption. In developed communities, dietary B_6 is fully adequate for normal nutrition.

NUTRITIONAL B_6 DEFICIENCY

The clinical effects of nutritional B_6 deficiency have been observed in volunteers subjected to vitamin restriction and in patients who have accidentally received a deficient diet. B_6 deficiency gives rise to anaemia, skin changes, sore tongue, peripheral neuropathy and convulsions[3][4][7]. The anaemia is of the so-called sideroblastic type. The red cells are generally small and hypochromic but the serum and marrow iron are increased.

Dietary B_6 deficiency is rare and though anaemia and convulsions have been recorded in infants on inadequate commercial milk formulae, the effects listed generally occur only when deficiency is secondary to other disease. Biochemical evidence of B_6 deficiency is, however, frequent. Signs of deficiency may not appear, but disordered tissue metabolism reversed by B_6 can be demonstrated, and the precipitating disorder may be improved by B_6 therapy.

LABORATORY DIAGNOSIS OF B_6 DEFICIENCY

Laboratory confirmation of suspected B_6 deficiency may be obtained by direct measurement of the concentration of the vitamin in plasma or red cells, or of its excretion product, 4-pyridoxic acid, in the urine[8][9]. The plasma vitamin may be measured by microbiological, chemical or enzymatic assay and low levels provide a good indication of depleted body vitamin stores. The level of 4-pyridoxic acid in the urine is usually measured by fluorometry. Decreased levels reflect low current dietary B_6 intake rather than reduced vitamin stores.

The microbiological assay for B_6 utilizes the fact that the vitamin is a growth factor for various yeasts and bacteria. By the appropriate choice of organism the different forms of the vitamin may be measured individually or collectively. Chemical assay can also measure each individual fraction. Enzymological assay measures only pyridoxal phosphate. For this a B_6 dependent enzyme, such as tyrosine decarboxylase[5] is treated to remove its coenzyme, leaving an inactive incomplete or apoenzyme. Sample addition restores activity in proportion to sample pyridoxal phosphate coenzyme concentration.

In addition to direct measurement of the vitamin concentration, B_6 deficiency can be diagnosed by its functional effects[9]. A common procedure is the tryptophan loading test[10][11]. B_6 is required for the conversion of tryptophan to nicotinic acid. If deficiency exists, this conversion is blocked, and there is an increased urinary excretion of the metabolite xanthurenic acid, and an increased hydroxykynurenine: hydroxyanthranilate ratio after oral tryptophan[12].

A second test of impaired function from B_6 deficiency is the demonstration of reduced red cell aspartate transaminase activity (AST)[13][14]. This results from diminished red cell concentration of coenzyme pyridoxal phosphate. Even more sensitive is the red cell AST activation test[15]. If pyridoxal phosphate is added to normal red cells, AST activity may approximately double[16], but if added to vitamin-deficient red cells, activation is even more marked. The higher the activation the greater the red cell B_6 deficiency.

The erythrocyte AST activation test can be recommended as a particularly convenient test for vitamin B_6 deficiency. AST determination is easy and inexpensive, and the enzyme is measured in all routine chemistry laboratories. The erythrocyte constitutes a readily available tissue, the method assesses tissue vitamin reserve rather than circulating dietary vitamin, it is sensitive, and it is specific for pyridoxal phosphate deficiency.

CONDITIONS GIVING B_6 DEFICIENCY

A wide variety of conditions give rise to biochemical vitamin B_6 deficiency, with low blood vitamin levels and abnormal functional tests. Metabolic abnormalities may be reversed by vitamin therapy, and such treatment improves the precipitating clinical disorder and any associated deficiency symptoms.

Deficient intake

This may result from abnormal diet, from age or from alcoholism. Biochemical deficiency is common in old age and has been reported in some 20% of subjects above the age of 65[17][18]. Although apparently symptomless, a DHSS nutrition survey has shown a strong positive correlation between B_6 deficiency and mortality in old age[19].

Impaired absorption

This generally results from gastrointestinal disease. Defective absorption may also contribute to deficiency in old age.

Increased demand

A major cause of B_6 deficiency is increased requirement for the vitamin. This may be physiological during growth, pregnancy and lactation, it is found in some 20% of patients on oral contraceptives[20], in hyperemesis gravidarum[21] or can result from increased catabolism as in thyrotoxicosis[22] and burns[23]. In the hyperactivity syndrome[24] and in epilepsy there may be an increased cerebral requirement for the vitamin. Deficiency may result in impaired brain formation of the inhibitory neurotransmitter γ-aminobutyric acid, lack of which may contribute to hyperactivity and convulsions. B_6 deficiency is present in some 30% of epileptics and B_6 therapy is beneficial in those who show biochemical evidence of deficiency[25].

The B_6 deficiency in patients receiving oral contraceptives, or in pregnancy, deserves special consideration because such deficiency may result in depression or impaired carbohydrate tolerance[20 26]. Oestrogens induce the enzymes which metabolize tryptophan and so increase the demand for pyridoxal phosphate. A relative deficiency of the vitamin results. One consequence of this is that brain hydroxytryptophan conversion to 5-hydroxytryptamine (serotonin) may be impaired. It is postulated that deficiency of brain serotonin gives rise to the depression. A second consequence of B_6 deficiency is impaired conversion of hydroxykynurenine to quinolinic acid. This latter normally inhibits excess glucose formation. When its concentration is reduced carbohydrate intolerance may result. Some 50% of patients with depression from oral contraceptives[27] and 50% of patients with pregnancy diabetes[28] show B_6 deficiency. Vitamin B_6 therapy of these patients in a dose sufficient to compensate for the increased metabolic demand, results in clinical improvement.

It should be noted that in patients on oestrogens or during pregnancy, the increased tryptophan metabolism raises xanthurenic acid excretion. In these patients, therefore, this cannot be used as a reliable test for B_6 deficiency.

Abnormal B_6 metabolism

Abnormal B_6 metabolism has been reported in patients on certain drugs, in alcoholics, in patients with liver disease, in pre-eclampsia and in asthma. Drug therapy with isonicotinic acid hydrazide (INH)[29 30], with L-dopa[31], hydrallazine or D-penicillamine[32] may all give rise to deficiency, usually as a result of complexing of the drug to the vitamin to yield an inactive metabolite. INH also inhibits the enzymes involved in pyridoxal phosphate synthesis. In the early days of high dose INH therapy for tuberculosis, peripheral neuritis from B_6 deficiency was a common complication. With conventional doses, neuropathy occurs in 3% of patients not given supplementary pyridoxal.

In alcoholics, low levels of B_6 have been reported in some 50% of patients[33], and seizure control during withdrawal may be improved by pyridoxal[34].

Superimposed on a deficient diet there is increased intracellular vitamin breakdown from ethanol metabolites[35]. It is important to be aware that the AST activation test is unsatisfactory for the diagnosis of vitamin B$_6$ deficiency in alcoholics. Activation is decreased, not increased as would be expected with deficiency[36], though the reason for this is unknown. In liver disease increased vitamin breakdown and diminished hepatic synthesis of active vitamin has also been suggested for the deficiency seen in up to 70% of patients[37][38]. Deficient placental vitamin synthesis may be a factor in pre-eclampsia[39]. In asthma[40] reduced formation of vasodilator bioamines may contribute to bronchospasm.

Increased excretion

Haemodialysis may result in B$_6$ deficiency[41], in part from increased elimination.

B$_6$ dependency states

The B$_6$ dependency states comprise a group of rare genetically determined disorders improved by pharmacological doses of the vitamin. Treatment may require vitamin intakes 100 times in excess of the normal daily requirement. In all these disorders there is inadequate function of an enzyme for which pyridoxal phosphate is a coenzyme. This may be due to defective apoenzyme synthesis, abnormal coenzyme binding or increased demand. The administration of high dose coenzyme provides compensatory stimulation or synthesis of the enzyme.

In infantile convulsions[42] impaired glutamate decarboxylase activity results in defective brain neurotransmitter synthesis and convulsions. In B$_6$ responsive anaemia[43][44] δ -aminolaevulinic acid synthase abnormality gives defective haem synthesis. Homocystinuria[45][46], cystathioninuria[47][48] and xanthurenic aciduria, all with mental retardation, may be a consequence of abnormality of the respective enzymes cystathionine synthase, cystathionase and kynurenase, and increased activity of glutamate–glyoxylate transaminase may be required to prevent renal stones and renal failure in primary hyperoxaluria[50].

EFFECT OF B$_6$ DEFICIENCY ON OTHER DIAGNOSTIC TESTS

Of importance to clinical biochemists is the fact that vitamin B$_6$ deficiency may interfere with performance of other diagnostic laboratory tests. One test in common use is serum aspartate transaminase determination for the diagnosis of myocardial infarction. The enzyme is liberated from the damaged heart and its increase in the serum can confirm infarction. B$_6$ deficiency may lower serum AST activity and could interfere with its diagnostic value[22]. However, adding pyridoxal to the assay system[51][52] compensates for dietary variation or deficiency.

CONCLUSION

In the introduction to this review it was stated that vitamin B_6 was unique because of the number and diversity of the metabolic reactions in which it is involved and the number of diseases in which its deficiency has been reported. This review of its clinical biochemistry has aimed to justify that claim.

References

1. Lumeng, L., Brashear, R. E. and Li, T-K. (1974). Pyridoxal 5-phosphate in plasma: source, protein-binding and cellular transport. *J. Lab. Clin. Med.*, **84**, 334
2. Reddy, S. K., Reynolds, M. S. and Price, J. M. (1958). The determination of 4-pyridoxic acid in human urine. *J. Biol. Chem.*, **233**, 691
3. Harper, H. H., Rodwell, V. W. and Mayes P. A. (1977). *Review of Physiological Chemistry*, 16th Edition, pp. 161–165. (Los Altos: Lange Medical Publications)
4. Marks, J. (1975). *A Guide to the Vitamins: Their Role in Health and Disease.* (Lancaster: MTP)
5. Boxer, G. E., Pruss, M. P. and Goodhar, R. S. (1957). Pyridoxal-5-phosphoric acid in whole blood and isolated leukocytes of man and animals. *J. Nutr.*, **63**, 623
6. *Recommended Dietary Allowances (1974).* 8th Edition. (Washington: National Academy of Science)
7. Coursin, D. B. (1961). Present status of vitamin B_6 metabolism. *Am. J. Clin. Nutr.*, **9**, 304
8. Henry, R. J. (1964). *Clinical Chemistry: Principles and Techniques*, pp. 723–724. (New York: Harper and Row)
9. Sauberlich, H. E., Canham, J. E., Baker, E. M., Raica, N. and Herman, Y. F. (1972). Biochemical assessment of the nutritional status of vitamin B_6 in the human. *Am. J. Clin. Nutr.*, **25**, 629
10. Greenberg, L. D., Bohn, D. F., McGrath, H. and Rhinehart, J. F. (1949). Xanthurenic acid excretion in the human subject on a pyridoxine-deficient diet. *Arch. Biochem.*, **21**, 237
11. Glazer, H. S. (1951). A study of urinary exretion of xanthurenic acid and other tryptophan metabolites in human beings with pyridoxine deficiency induced by desoxypyridoxine. *Arch. Biochem.*, **33**, 243
12. Heeley, A. F. (1965). The effect of pyridoxine on tryptophan metabolism in phenylketonuria. *Clin. Sci.*, **29**, 465
13. Marsh, M. E., Greenberg, L. D. and Rhinehart, J. F. (1955). The relationship between B_6 investigation and transaminase activity. *J. Nutr.*, **56**, 115
14. Sass, M. and Murphy, G. T. (1958). The effect of isonicotinic acid hydrazide and vitamin B_6 on glutamic–oxaloacetic transaminase levels in whole blood. *Am. J. Clin. Nutr.*, **6**, 424
15. Raica, N. and Sauberlich, H. E. (1964). Blood cell transaminase activity in human vitamin B_6 deficiency. *Am. J. Clin. Nutr.*, **15**, 67
16. Bayoumi, R. A. and Rosalki, S. B. (1976). Evaluation of methods of co-enzyme activation of erythrocyte enzymes for detection of deficiency of vitamins B_1, B_2 and B_6. *Clin. Chem.*, **22**, 327
17. Hamfelt, A. (1964). Age variation of vitamin B_6 metabolism in man. *Clin. Chim. Acta*, **10**, 48
18. Hoorn, R. K. J., Flikweert, J. P. and Westerink, D. (1975). Vitamin B-1, B-2 and B-6 deficiencies in geriatric patients, measured by co-enzyme stimulation of enzyme activities. *Clin. Chim. Acta*, **61**, 151
19. Hodkinson, H. M. and Exton-Smith, A. N. (1976). Factors predicting mortality in the elderly in the community. *Age Ageing*, **5**, 110
20. Rose, D. P., Strong, R., Adams, P. W. and Harding, P. E. (1972). Experimental vitamin B_6 deficiency and the effect of oestrogen-containing oral contraceptives on tryptophan metabolism and vitamin B_6 requirements. *Clin. Sci.*, **42**, 465
21. Gant, H., Reinken, L., Dapant, O and Scholz, K. (1975). Vitamin B_6 — Verarmung bei hyperemesis gravidarum. *Wien. Klin. Wochenschr.*, **87**, 510
22. Hamfelt, A. (1966). The effect of pyridoxal phosphate on the aminotransferase assay in blood. *Scand. J. Clin. Lab. Invest.*, **18** (Suppl. 92), 181

23. Barlow, G. B. and Wilkinson, A. W. (1975). Plasma pyridoxal phosphate metabolism in children with burns and scalds. *Clin. Chim. Acta*, **64**, 79

24. Bhagavan, H. N., Coleman, M. and Coursin, D. B. (1975). The effect of pyridoxine hydrochloride on blood serotonin and pyridoxal phosphate contents in hyperactive children. *Pediatrics, (Springfield)*, **55**, 437

25. Hagberg, B., Hamfelt, A. and Hansson, O. (1966). Tryptophan load tests and pyridoxal-5-phosphate levels in epileptic children. *Acta Paediat. Scand.*, **55**, 371

26. Wynn, V. (1975). Vitamins and oral contraceptive use. *Lancet*, **i**, 561

27. Adams, P. W., Wynn, V., Rose, D. P., Seed, M. Folkard, J. and Strong, R. (1973). The effect of pyridoxine hydrochloride (vitamin B_6) upon depression associated with oral contraception. *Lancet*, **i**, 897

28. Bennink, H. J. T. C. and Schreurs W. H. P. (1975). Improvement of oral glucose tolerance in gestational diabetes by pyridoxine. *Br. Med. J.*, **3**, 13

29. Biehl, J. P. and Vilter, R. W. (1954). Effect of isoniazid in vitamin B_6 metabolism: its possible significance in producing isoniazid neuritis. *Proc. Soc. Exp. Biol. Med.*, **85**, 389

30. Biehl, J. B. and Vimitz, H. J. (1954). Studies on the use of a high dose of isoniazid. *Am. Rev. Tuberc.*, **70**, 430

31. Evered, D. F. (1971). L-dopa as a vitamin B_6 antagonist. *Lancet*, **i**, 914

32. Hollister, L. E., Moore, F. F., Forrest, F. and Bennett, J. L. (1966). Antipyridoxine effect of D-penicillamine in schizophrenic men. *Am. J. Clin. Nutr.*, **19**, 307

33. Davis, R. E. and Smith, B. K. (1974). Pyridoxal and folate deficiency in alcoholics. *Med. J. Aust.*, **2**, 357

34. Lunde, F. (1960). Pyridoxine deficiency in chronic alcoholism. *J. Nerv. Ment. Dis.*, **131**, 77

35. Veitch, R. L., Lumeng, L. and Li, T-K. (1975). Vitamin B_6 metabolism in chronic alcohol abuse. *J. Clin. Invest.*, **55**, 1026

36. Rosalki, S. B. (1977). Enzyme tests for alcoholism. *Rev. Epidém. Santé Publ.*, **25**, 147

37. Mitchell, D., Wagner, C., Stone, W. J., Wilkinson, G. R. and Schenker, S. (1976). Abnormal regulation of plasma pyridoxal-5-phosphate in patients with liver disease. *Gastroenterology*, **71**, 1043

38. Labadarios, D., Rossouw, J. E., McConnell, J. B., Davis, M. and Williams, R. (1977). Vitamin B_6 deficiency in chronic liver disease — evidence for increased degradation of pyridoxal-5-phosphate. *Gut*, **18**, 23

39. Gaynor, R. and Dempsey, W. B. (1972). Vitamin B_6 enzymes in normal and pre-eclamptic human placentae. *Clin. Chim. Acta*, **37**, 411

40. Collipp, P. J., Glodzier, S., Weiss, N., Soleymani, Y. and Snyder, R. (1975). Pyridoxine treatment of childhood bronchial asthma. *Ann. Allergy*, **35**, 93

41. Stone, W. J., Warnock, L. G. and Wagner, C. (1975). Vitamin B_6 deficiency in uremia. *Am. J. Clin. Nutr.*, **28**, 950

42. Hunt, A. D. Jr., Stokes, J., McCrory, W. W. and Stroud, H. H. (1954). Pyridoxine dependency: a report of a case of intractable convulsions in an infant controlled by pyridoxine. *Pediatrics, (Springfield)*, **13**, 140

43. Harris, J. W., Whittington, R. M., Weisman, R. and Horrigan, D. L. (1956). Pyridoxine responsive anaemia in the human adult. *Proc. Soc. Exp. Biol. Med.*, **91**, 427

44. Horrigan, D. L. and Harris, J. W. (1964). Pyridoxine-responsive anaemia. *Adv. Intern. Med.*, **12**, 103

45. Barber, G. W. and Spaeth, G. L. (1967). Pyridoxine therapy in homocystinuria. *Lancet*, **i**, 1384

46. Mudd, S. H., Edwards, W. A., Loeb, P. M., Brown, M. S. and Laster, L. (1970). Homocystinuria due to cystathionine synthase deficiency: the effect of pyridoxine. *J. Clin. Invest.*, **49**, 1762

47. Frimpter, G. W., Haymovitz, A. and Horwith, M. (1963). Cystathioninuria. *N. Engl. J. Med.*, **268**, 333

48. Shaw, K. N. F., Lieberman, E., Koch, R., Donnell, G. N. (1967). Cystathioninuria. *Am. J. Dis. Child.*, **113**, 119

49. Tada, K., Yokoyama, Y., Nakagawa, H., Jo Shida, T. and Askawa, T. (1967). Vitamin B_6 dependent xanthurenic aciduria. *Tohoku J. Creper. Med.*, **93**, 1157

50. Gibbs, D. A. and Watts, R. W. E. (1970). The action of pyridoxine in primary hyperoxaluria. *Clin. Sci.*, **38**, 277

51. Rej, R., Fasce, C. F. and Vanderlinde, R. E. (1973). Increased aspartate aminotransferase activity of serum after in vitro supplementation with pyridoxal phosphate. *Clin. Chem.*, **19**, 92
52. Rosalki, S. B. and Bayoumi, R. A. (1975). Activation by pyridoxal phosphate of aspartate transaminase in serum of patients with heart and liver disease. *Clin. Chim. Acta*, **59**, 357

8

Inappropriate vitamin C reserves: their frequency and significance in an urban population

C. J. SCHORAH

INTRODUCTION

The introduction of the potato into Europe in the 16th century, and later the use of fruit juice on board ship during the Napoleonic wars, led to a rapid decline in the incidence of scurvy so that it is now a relatively rare disease in the UK. Yet interest in the antiscorbutic agent, vitamin C, persists. Essentially, interest and controversy exist in three areas. The first concerns the function of vitamin C. Vitamin C is involved in the hydroxylation of proline during collagen synthesis[2], but it is unlikely that this metabolic function can explain all the symptoms of scurvy, or that it is the major reason for death in the scorbutic patient. A number of other possible roles of ascorbic acid have been suggested. It is believed to protect tetrahydrofolic acid from oxidation to folic acid[3 4], to promote the absorption of iron[5 6], and to aid liver microsomal drug metabolism[7 8]. It is questionable, however, whether vitamin C is essential for these functions and it is possible therefore that a major role of vitamin C metabolism has yet to be discovered.

The second area of controversy is in the possible role of megadose vitamin C therapy (>500 mg/d) in maintaining health and preventing or reducing the incidence of diseases such as cancer, arterial disease and infection[9-11]. There is evidence that vitamin C can promote the immune response[11-13], and possibly, therefore, aid in the treatment and prevention of malignancy and infection. There is also evidence that conversion of cholesterol to bile acids is increased by vitamin C[14], a response which could reduce serum cholesterol[15]. However, controlled double-blind trials have yet to be reported which investigate the effects of vitamin C treatment on tumour growth of the progression of arterial disease, whilst such trials have indicated that large doses of vitamin C are unable either to prevent or cure the common cold[16 17].

Although the need for a large intake of vitamin C has yet to be proved, there is clear evidence that the intakes of a proportion of the population of the UK, rather than being high, are below the minimum recommended by the Department of Health and Social Security (30 mg/d)[18-22]. There is cause for concern because some sections of the community who have these low intakes, such as the poor during pregnancy and patients in hospital, arguably require larger reserves in order to facilitate the growth or repair of tissue[22 24], to combat infection[25] and to aid drug metabolism[8 11]. However, although intake has been assessed, with the exception of elderly subjects[18 26 27], little is known about the cellular levels of vitamin C in these groups and almost no work has been directed at assessing the benefit of vitamin C supplements. This is a third area of interest in vitamin C and the one that I wish to consider in more detail.

VITAMIN C RESERVES IN SELECTED POPULATIONS

An assessment of vitamin C status requires a definition of what, in the light of current evidence, is an appropriate tissue vitamin C concentration. Estimates of the appropriateness of various leukocyte reserves are made in Table 1. The leukocyte measure is strictly an assessment of both platelet and leukocyte vitamin C and it is believed to reflect the vitamin C status of tissues more accurately than plasma[28 29]. The data in Table 1 have been summarized from records of dietary intakes and the levels of vitamin C that they maintain in the plasma and in the leukocyte, and the information may not accord exactly with any one author.

Table 1 Intakes of vitamin C, resulting plasma and leukocyte vitamin C levels and suggestions on the adequacy of these reserves

Group	Intake (mg)	Plasma (μg/ml)	Leukocytes (μg/10^8)	Tissue status
1	>100	>10	40–50	Saturated
2	25–100	3.5–10	20–40	Adequate
3	10–25	1.5–3.5	8–20	Inadequate
4	<10	<1.5	<8	Deficient

In group 1, intakes maintain tissue saturation. The need for tissue saturation with vitamin C has yet to be proved and hence the tissue reserves of group 2, although lower than group 1, are considered adequate. The intakes of group 4 have been shown by a number of workers to be associated with clinical scurvy[30-33] and hence tissues are considered to be deficient. There is, however, some evidence that intakes and tissue reserves of group 3, whilst not producing frank clinical scurvy, can impair health. Effects that have been reported include depression and hysteria[34], reduced wound healing[29 35 36], decreased urine hydroxyproline[37] and increased urine proline[38]. Thus, current evidence suggests that leukocyte vitamin C concentrations in group 3 are inadequate.

The tentative upper limits for both deficient and inadequate leukocyte levels of vitamin C are marked in Figure 1 which illustrates the mean and 95% range for leukocyte vitamin C in a number of populations who are believed to have poor intakes. The pregnant population consists of results collected during the first trimester of pregnancy on over a thousand pregnant women. The institutionalized subjects were a geriatric group older than 65 years and a younger population with an age range 24–54 years. All the institutionalized patients had been in hospital for more than six months. Leukocyte vitamin C was also measured in a group of patients prior to major surgery. A control group of blood donors aged 19–55 was selected to compare with the hospital patients.

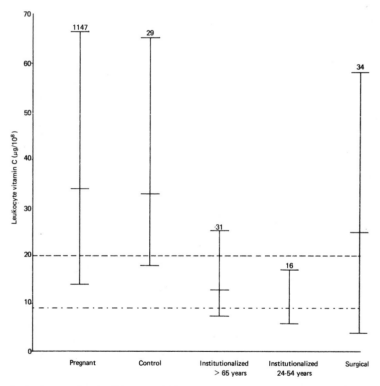

Figure 1 Leukocyte vitamin C (mean and 95% range) in different populations and number of subjects in each. _____ Upper limit of leukocyte vitamin C when reserves are inadequate; _._._ Upper limit of leukocyte vitamin C when reserves are deficient

It is clear that the pregnant and control groups have similar vitamin C distributions but both the old and the younger institutionalized subjects have a much lower level of cellular vitamin C. Results in surgical patients are widely distributed but a proportion have levels lower than the control group. The low vitamin C reserves in the institutionalized were confirmed by the mean

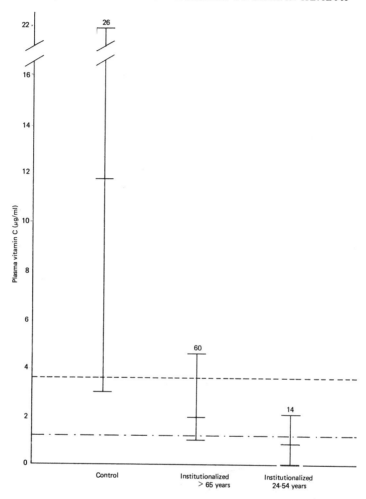

Figure 2 Plasma vitamin C (mean and 95% range) in different populations and number of subjects in each. ____ Upper limit of plasma vitamin C when reserves are inadequate; _._._ Upper limit of plasma vitamin C when reserves are deficient

and 95% range for plasma vitamin C (Figure 2). In terms of the adequacy of vitamin C status in these populations, none of the subjects had clinical scurvy although some of the institutionalized were bruised. The majority of the pregnant population had adequate reserves of vitamin C, only 9.5% of women having values below the upper limit of the inadequate zone ($<20\,\mu g/10^8$ leukocytes). However, the fact that women in this group were pregnant complicates a comparison with a general population. Higher reserves may be appropriate to ensure an adequate supply to the rapidly developing fetus. In addition, decreased vitamin C levels in this large population were associated with increasing social class number, smoking and the spring season[39]. These factors appeared to affect the leukocyte vitamin C levels independently. This is illustrated in Table 2 where the population groups identified by extremes of class, season and smoking have almost separate vitamin C distributions. Hence many women who smoke, are poor, and who are pregnant during spring have a leukocyte vitamin C level which would be considered unsatisfactory in the general population. The effect of smoking on leukocyte vitamin C is unknown, but the seasonal changes could be caused by consumption of old potatoes in April and early May, a time when the vitamin C content of the potato is considerably reduced[40]. The social class difference is consistent with the consumption of inadequate diets by the poor.

Table 2   Leukocyte vitamin C levels in extreme groupings of pregnant women

Class*	Season	No. of cigarettes/d	No. of women	Vitamin C ($\mu g/10^8$ leukocytes)		
				Mean	S.D.	95% range
I + II	July, August, September	0	33	45.1	12.1	29–70
IV + V	April, May	10–20	11	21.7	6.6	14–34

* Based on the Registrar General's classification[42] (Office of Population Censuses and Surveys, 1970): I, professional; II, intermediate occupations; IV, partly skilled; V, unskilled

The leukocyte reserves in the institutionalized groups are unsatisfactory, several reaching levels associated with clinical scurvy. This could be due to hospital diet, the ascorbic acid content of which is reported to be poor[20]. A number in the surgical group also have ascorbic acid reserves close to those found in clinical scurvy, a factor which could be important because of the need for wound healing in this population. However, although there is an impression that the effects of reduced vitamin C levels may be detrimental even in the absence of clinical scurvy, there has been little research directed at investigating whether this is the case.

POSSIBLE EFFECTS OF INADEQUATE VITAMIN C RESERVES

One might assume that a vitamin C depletion in the first trimester of pregnancy, if of consequence, would influence the early events of pregnancy. It is of interest therefore, that a similar range of leukocyte vitamin C ($12–33.5 \mu g/10^8$ leukocytes) to that of the disadvantaged group in Table 2 was seen in five mothers who subsequently gave birth to children with neural tube defects (e.g. spina bifida or anencephaly), lesions which almost certainly develop in the first 30 days of pregnancy. The mothers in this neural tube defect group also had a lower level of blood folate and serum B_{12}[41]. There is a relationship between these vitamins in the metabolism of purines and pyrimidines and hence in maintaining the rate of DNA synthesis, cell division and neural tube closure. Folic acid and B_{12} are closely involved and vitamin C

Table 3 Observations on birthweight (% of mothers with children < or > 3250 g birthweight who had a leukocyte vitamin C<20 μg/10⁸l, a leukocyte vitamin C<16μg/10⁸l, or who were from social class IV or V† or smoked > 10 cigarettes/d)

	Birthweight <3250 g (% total)	Birthweight >3250 g (% total)
Vitamin C (<20 μg/10⁸l)	***14.3	*** 6.7
Vitamin C (<16 μg/10⁸l)	* 5.1	* 1.8
Social class IV and V†	16.4	13.8
Smoking≥10 cigarettes/d	**26.3	**16.3

* $2 \times 2 x^2$ analyses were significant: $*p<0.05$, $**p<0.005$, $***p<0.001$
†   Based on the Registrar General's classification[42] (Office of Population Censuses and Surveys, 1970): IV, partly skilled; V, unskilled

probably maintains folic acid in the reduced (active) form[3 4]. It is possible that reduced levels of these vitamins acting together could contribute to the causation of neural tube defects.

Nutrition near to term will probably be important in determining the size of the child. Nevertheless there is an increased incidence of low first trimester vitamin C values in mothers of lighter birthweight children (Table 3). This surprisingly enough does not appear to be related to social class for the distribution of the lower social class groups between the two birthweight categories in Table 3 is similar. Smoking, however, is increased in the group of mothers who give birth to children lighter than 3250 g and maternal smoking during pregnancy is known to be associated with reduced birthweight. However, as lower leukocyte vitamin C is observed in smokers, an increased incidence of smoking in the lower birthweight group does not necessarily argue against vitamin C as a factor in determining birthweight.

In surgical patients the lowest vitamin C reserves are seen in those patients who have been in hospital for more than a week after major surgery, at a time when many of these patients had sepsis (Table 4). It is possible to suggest a link between these low vitamin C values and the poor recovery of these patients, but this association and that of vitamin C to birthweight and the incidence of neural tube defects could have arisen by chance and does not necessarily imply a cause and effect relationship. This can only be evaluated by testing if vitamin C therapy can improve health in these groups of subjects. Surgical and neural tube defect groups have inappropriate levels of other vitamins in addition to vitamin C and trials of multivite preparations are underway in these groups. But by far the most depleted vitamin in the geriatric group is vitamin C and hence we have carried out a double blind trial on the effect of improved vitamin C intake on the clinical state and bodyweight of long-stay elderly patients in geriatric hospitals.

Table 4 Vitamin C levels in surgical patients compared with controls*

Category and number in each	Mean vitamin C (μg/ 10^8 leukocytes)	Frequency of low values (<lower limit of 95% range in controls)(%)
Controls* (21)	20.1	5
Before major surgery (8)	22.5	25
<1 week after major surgery (11)	17.5	18
>1 week after major surgery (11)	12.5	46

* Controls were laboratory staff, hospital employees or patients who presented at the hospital with a minor problem

CHANGES IN HEALTH WHEN VITAMIN C INTAKES ARE INCREASED

Table 5 shows that vitamin C levels of the 115 patients in the trial were well below those in a younger population and were also lower than subjects of a similar age range outside hospital. Indeed the values overlapped the range found in frank clinical scurvy.

Table 5 Initial vitamin C values in the institutionalized geriatric population compared with younger subjects and patients with scurvy

Subjects	Age range (years)		Vitamin C				
			(μg/ 10^8 leukocytes)			plasma (μg/ml)	
		n	Mean	95% range	n	Mean	95% range
Young subjects	19–55	46	31.5	16.2–54	31	10.3	3.1–21
Elderly outpatients	63–84	27	18.7	8.9–36.5	27	4.6	1.7–18.3
Geriatric patients (institutionalized)	59–97	42	11.2	4.2–24.7	111	1.9	0.7–5.1
Patients with clinical scurvy*	—	30	4.0	0.0–8.0	34	0.8	0.0–1.5

Mean values for both plasma and leukocyte vitamin C in the geriatric patients were significantly lower than mean values for the young subjects ($p < 0.001$) and elderly outpatients ($p < 0.02$)

* See references 30–33

Patients were assessed clinically and were also weighed before supplementation. The patients were then randomly allocated to either placebo or vitamin C (1 g/d) groups, the treatment being maintained for 28 days. At the end of the trial the patients were reweighed and assessed for any changes in

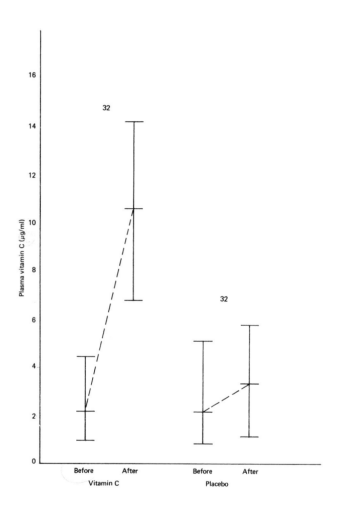

Figure 3 Plasma vitamin C (mean and 95% range) in geriatric patients before and after supplementation with 1 g vitamin C/d or placebo for one month, and the number of estimations in each group

living activities, such as mobility, dressing, feeding and social conversation, interest in surroundings or appetite. A number of patients also had plasma and leukocyte vitamin C remeasured.

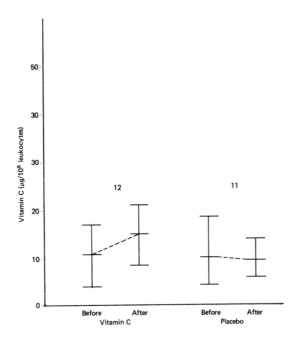

Figure 4 Leukocyte vitamin C (mean and 95% range) in geriatric patients before and after supplementation with 1 g vitamin C/d or placebo for one month, and the number of estimations in each group

Figure 3 shows that marked increases had occurred in the plasma vitamin C in those subjects on supplements during the trial and in addition there was also a slight but significant increase in the placebo group. Leukocyte vitamin C also increased significantly in the supplemented patients, but this increase was slight compared with the dose given (Figure 4). It could be expected that

It could be expected that no dramatic clinical improvement would be seen in many of the subjects, most of whom had been in hospital for many months for a variety of reasons and at the end of the trial the majority of the patients had not changed by any assessment. However, the numbers of improvements were greater and the numbers of deteriorations less in the vitamin C group compared with the placebo ($p<0.025$, x^2 analysis; Table 6). Weight changes were also significantly different between the two groups (Table 7) increasing in the vitamin C group and decreasing in the placebo and this difference was more apparent in the males. These advantages seen in the supplemented group suggest that vitamin C may be effective in improving health in these patients. The difference between the groups might have been greater if the vitamin supplements had returned leukocyte vitamin C to normal at the end of the trial.

Table 6 Clinical changes in institutionalized geriatric subjects following one month of vitamin C or placebo supplementation

	Vitamin C				Placebo			
	++	+	0	—	++	+	0	—
Subjects with one or more change*	1	14	34	7	0	8	40	11
Number of changes*	3	23	132	10	0	15	143	19

++ Improved considerably; + improved slightly; 0 no change; — deteriorated slightly
* Changes were in daily living activities, interest in surroundings and appetite
In the vitamin C group the numbers of overall improvements compared (a) with deteriorations and (b) with deterioration plus no change were significantly greater than in the placebo group. (X^2 analysis; $p<0.025$; $p<0.05$ respectively)

Table 7 Weight changes in geriatric patients during vitamin C or placebo supplementation

Subjects	Weight change (kg) (mean $\pm$ SE)	
	Ascorbic acid	Placebo
Total population	+0.54* $\pm$0.29	—0.24 $\pm$ 0.25
Males	+0.74** $\pm$0.29	—0.66 $\pm$ 0.50

Significantly different from placebo; *$p<0.05$; **$p<0.025$

CONCLUSIONS

Irrespective of the suggested need for large doses of vitamin C in certain conditions, it is clear that a proportion of the population, such as the poor during pregnancy, some patients awaiting major surgery and those in hospital for long periods, have poor vitamin C reserves possibly because their intakes are below the recommended. There is an impression that these reserves, whilst not causing clinical scurvy, may be suboptimal. They are reported to be associated with impaired collagen metabolism, decreased wound healing and neurological change and we find that they are more frequent in women who give birth to smaller babies or neural tube defect infants and are associated with poor recovery from surgical procedures. Vitamin C supplementation in geriatric patients indicates that these reserves are indeed, at any rate for this group, suboptimal and suggest that further investigations, in addition to confirming this observation, should be directed at the surgical and young institutionalized groups to determine if better vitamin C nutrition can improve health and aid recovery from disease and trauma in these groups.

Acknowledgements

This work was completed in association with Professor D. B. Morgan,

Professor R. W. Smithells, Mr G. L. Hill, Mrs A. Newill and Dr D. Scott. We thank all patients involved in the studies and the nursing staff and pharmacists who did much of the work in the geriatric trial. We gratefully acknowledge generous financial support from Action for the Crippled Child and Roche Products Ltd.

References

1. Wilson, C. G. (1975). The clinical definition of scurvy and the discovery of vitamin C. *J. Hist. Med.*, **30**, 40
2. Barnes, M. J. (1975). The function of ascorbic acid in collagen metabolism. *Ann. N.Y. Acad. Sci.*, **258**, 264
3. Banergee, S. and Nandy, N. (1970). Pteroylglumatic acid nutrition in vitamin C deficient guinea pigs. *Proc. Soc. Exp. Biol. Med.*, **133**, 151
4. Stokes, P. L., Melikian, V., Lemming, R. L., Portman-Graham, H., Blair, J. A. and Cooke, W. T. (1975). Folate metabolism in scurvy. *Am. J. Clin. Nutr.*, **28**, 126
5. Prasad, J. S. (1975). Effect of ascorbic acid on plasma iron turnover. *Clin. Chim. Acta*, **59**, 101
6. Bingöl, A., Altay, C., Say, B. and Donmez, S. (1975). Plasma, erythrocyte and leukocyte ascorbic acid concentrations in children with iron deficiency anaemia. *J. Pediatr.*, **86**, 902
7. Street, J. C. and Chadwick, R. W. (1975). Ascorbic acid requirements and metabolism in relation to organochlorine pesticides. *Ann. N.Y. Acad. Sci.*, **258**, 132
8. Zannoni, V. G. and Sato, P. H. (1975). Effects of ascorbic acid on microsomal drug metabolism. *Ann. N.Y. Acad. Sci.*, **258**, 119
9. Cameron, E. and Pauling, L. (1978). Supplemental ascorbate in the supportive treatment of cancer. *Proc. Natl. Acad. Sci. USA.*, **75**, 4538
10. Loh, H. S. (1973). Mortality from atherosclerosis and vitamin C intake. *Lancet*, **ii**, 153
11. Wilson, C. W. M. (1975). Clinical pharmacological aspects of ascorbic acid. *Ann. N.Y. Acad. Sci.*, **258**, 355
12. Siegel, B. V. and Morton, J. I. (1977). Vitamin C and the immune response. *Experientia*, **33**, 393
13. Thomas, W. R. and Holt, P. G. (1978). Vitamin C and immunity: an assessment of the evidence. *Clin. Exp. Immunol.*, **32**, 370
14. Ginter, E. (1975). Ascorbic acid in cholesterol and bile acid metabolism. *Ann. N.Y. Acad. Sci.*, **258**, 410
15. Ginter, E. (1976). Vitamin C and plasma lipids. *N. Engl. J. Med.*, **294**, 559
16. Coulehan, J. L., Eberhard, S., Kapner, L. and Taylor, F. (1976). Vitamin C and acute illness in Navajo schoolchildren. *N. Engl. J. Med.*, **295**, 973
17. Miller, J. Z., Nance, W. E., Norton, J. A., Wolen, R. L., Griffith, R. S. and Rose, R. J. (1977). The therapeutic effect of vitamin C. *J. Am. Med. Assoc.*, **237**, 3248
18. Andrews, J., Brook, M. and Allen, M. A. (1966). Influence of abode and season on the vitamin C status of the elderly. *Geront. Clin.*, **8**, 257
19. Allen, R. J. L., Brook, M. and Broadbent, S. R. (1968). The variability of vitamin C in our diet. *Br. J. Nutr.*, **22**, 555
20. Eddy, T. P. (1968). The problem of retaining vitamins in hospital food. In: A. N. Exton-Smith and D. L. Scott (eds.). *Vitamins in the Elderly*, pp. 86–92. (Bristol: John Wright & Sons)
21. Smithells, R. W., Ankers, C., Carver, M. E., Lennon, D., Schorah, C. J. and Sheppard, S. (1977). Maternal nutrition in early pregnancy. *Br. J. Nutr.*, **38**, 497
22. Department of Health and Social Security. (1969). *Rep. Publ. Hlth. Med. Subj., Lond.*, **120**
23. Chen, T. L. and Raisz, L. G. (1975). The effects of ascorbic acid deficiency on calcium and collagen metabolism in cultured fetal rat bones. *Calcif. Tissue Res.*, **17**, 113
24. Hume, R., Weyers, E., Rowan, T., Reid, D. S. and Hillis, W. S. (1972). Leukocyte ascorbic acid levels after acute myocardial infarction. *Br. Heart J.*, **34**, 238
25. Hume, R. and Weyers, E. (1973). Changes in leukocyte ascorbic acid during the common cold. *Scott. Med. J.*, **18**, 3
26. Bowers, E. F. and Kubik, M. M. (1965). Vitamin C levels in old people and the response to ascorbic acid and to the juice of the acerola. *Br. J. Clin. Pract.*, **19**, 141

27. Burr, M. L., Elwood, P. C., Hole, D. J., Hurley, R. J. and Hughes, R. E. (1974). Plasma and leukocyte ascorbic acid levels in the elderly. *Am. J. Clin. Nut.*, **27**, 144
28. Andrews, J. and Brook, M. (1968). The relationship of plasma and leukocyte (including platelet) ascorbic acid content. *Geront. Clin.*, **10**, 128
29. Gerson, C. D. (1968). Ascorbic acid deficiency in clinical disease. *Ann. N. Y. Acad. Sci.*, **258**, 483
30. Goldberg, A. (1963). The anaemia of scurvy. *Q. J. Med.*, **32**, 51
31. Hodges, R. E., Hood, J., Canham, J. E., Sauberlich, H. E. and Baker, E. M. (1971). Clinical manifestations of ascorbic acid deficiency in man. *Am. J. Clin. Nut.*, **24**, 432
32. Bartley, W., Krebs, H. A. and O'Brien, J. R. P. (1953). The vitamin C requirements of human adults. *Spec. Rep. Ser. Med. Res. Coun.*, **280**
33. Crandon, J. H., Landau, B., Mikal, S., Balmanno, J., Jefferson, M. and Mahoney, N. (1958). Ascorbic acid economy in surgical patients as indicated by blood ascorbic acid levels. *N. Engl. J. Med.*, **258**, 105
34. Kinsman, R. A. and Hood, J. (1971). Some behavioral effects of ascorbic acid deficiency. *Am. J. Clin. Nutr.*, **24**, 455
35. Crandon, J. H., Lennihan, R. and Mikal, S. (1961). Ascorbic acid economy in surgical patients. *Ann. N. Y. Acad. Sci.*, **92**, 246
36. Taylor, T. V., Rimmer, S., Day, B., Butcher, J. and Dymock, I. W. (1974). Ascorbic acid supplementation in the treatment of pressure-sores. *Lancet*, **ii**, 544
37. Windsor, A. C. W. and Williams, C. B. (1970). Urinary hydroxyproline in the elderly with low leukocyte ascorbic acid levels. *Br. Med. J.*, **1**, 732
38. Bates, C. J. (1977). Proline and hydroxyproline excretion and vitamin C status in elderly human subjects. *Clin. Sci. Mol. Med.*, **52**, 535
39. Schorah, C. J., Zemroch, P. J., Sheppard, S. and Smithells, R. W. (1978). Leukocyte ascorbic acid and pregnancy. *Br. J. Nutr.*, **39**, 139
40. McCance, R. A. and Widdowson, E. M. (1960). *Spec. Rep. Ser. Med. Res. Coun.*, **297**, 199
41. Smithells, R. W., Sheppard, S. and Schorah, C. J. (1976). Vitamin deficiencies and neural tube defects. *Arch. Dis. Childh.*, **51**, 944
42. Office of Population Censuses and Surveys. (1970). *Classification of Occupations.* (London: HMSO)

9

Developments in analytical methods for the determination of fat-soluble vitamins in foods

R. A. WIGGINS

A large part of the time of an analyst concerned with the determination of the fat-soluble vitamins in foods is often spent in the chromatographic separation of vitamins from extracts prior to physicochemical measurements. It is not surprising, therefore, that developments in chromatography that improve what could be called the three 'Rs' of chromatography, i.e. resolution, reliability and rapidity of separation, are welcomed.

The application of gas chromatography led to some major advances in the analysis of vitamins D and E in foods. Methods for the gas chromatographic determination of these vitamins in foods have been developed at the Laboratory of the Government Chemist (LGC)[1] [2]. Gas chromatography has, however, not been applied to any great extent to the analysis of vitamin A because of thermal stability problems although it is possible to separate the silyl ethers of retinol by gas chromatography[3] [4].

Recent years, however, have seen the increasing application of high performance liquid chromatography (HPLC) to the analysis of fat-soluble vitamins. In many cases the use of HPLC has resulted in a noticeable improvement in all three 'Rs' but the particular merit of the application of this technique has been to increase the 'rapidity' of methods. The economics of the application of HPLC are continually improving as increased market competition has reduced prices. It is also relatively easy to pack one's own HPLC column using apparatus shown in Figure 1. This apparatus[5] consists of a stainless steel pot, manufactured to withstand internal pressures of 6000 p.s.i., which stands on a magnetic stirrer. A slurry of the HPLC packing is placed in the pot and is kept stirred by a magnetic follower. The empty HPLC column, fitted with suitable end fittings, is clamped to a port in the top of the stainless steel reservoir and a normal HPLC pump is connected to another port. The pump is then switched on (normally at its fastest pumping speed [approx

73

10 ml/min]) and the column packing is forced into the column. The whole procedure takes about 45 min and has been used successfully at LGC to pack columns with 5 micron packing materials.

Figure 1 Apparatus for packing HPLC columns

HPLC is currently used at the LGC for the analysis of vitamin A (retinol) and vitamin D in foods. Before application of HPLC, retinol was separated from food extracts by partition chromatography on conventional open columns which took 1½–2 h to develop[6]. The same separation can however be achieved in 10–15 min on an HPLC column of 5 micron reverse phase packing with methanol/water mixture as mobile phase. HPLC has thus reduced the time of analysis for retinol in foods from 3–4 h to 1–2 h. Figure 2 illustrates some HPLC separations of retinol from a number of foods using a method developed at LGC[7]. The HPLC system described in Figure 2 will also separate the all-*trans* and 13-*cis* isomers of retinol. (The 13-*cis* isomer has approximately 75% of the activity of all-*trans*-retinol.) A similar system[8] has been used to determine the 13-*cis*-retinol content of a number of foods. This study found the following proportions of 13-*cis*-retinol:– wheat cereal – 13%, butter – 18%, cod liver oil – 35%, dry pet food – 26%.

The increased resolution obtained using microparticulate column packing means that it is important to use specific nomenclature when reporting results e.g. specify that a result is for all-*trans*-retinol content and not just retinol. This is particularly important when results from various methods are being compared.

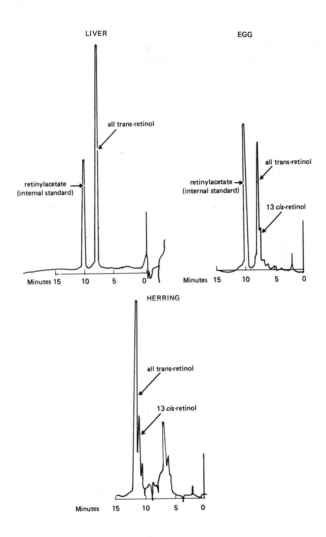

Figure 2 Separation of retinol from foods using HPLC. Column — 5 micron Spherisorb ODS; mobile phase — 90–95% methanol/10–5% water; detector — UV absorbance 313 nm

An analysis of the vitamin A activity in foods should also include the contribution made from those carotenoids which have provitamin A activity. There are more than 80 naturally occurring carotenoids of which about 10 have provitamin A activity, though for nutritional analysis interest is often restricted to the four with most activity, i.e. α-carotene, β-carotene, γ-carotene and cryptoxanthin. Separation of these carotenoids by conventional chromatography usually involves a crude separation on a column of magnesium oxide according to polarity followed by separation of individual carotenoids by thin layer chromatography. Recently columns of magnesium oxide have been used in pressurized chromatographic systems to separate α- and β-carotenes and cryptoxanthin[9]. An acetone-hexane gradient separated these three carotenoids in about 20 min. The separation has been applied quantitatively to the analysis of provitamin A activity in a number of citrus juices.

The first application of HPLC to the analysis of fat-soluble vitamins at LGC was to the analysis of vitamin D. Although a gas chromatographic method[1] had been developed the extensive extract purification required meant that the analysis took 2 days to complete. Application of HPLC has reduced the time of analysis to 5-6 h. The analysis time is still quite long because it was found that some preliminary purification of extracts was necessary before the vitamins D could be separated, sufficiently for quantitative analysis, by HPLC. The two nutritionally important forms of vitamin D, vitamin D_2 (ergocalciferol) and vitamin D_3 (cholecalciferol) can be separated by HPLC though so far this separation can only be achieved on 5 micron reverse phase packings[10]. These column packings will also separate the isotachysterols formed on treatment of the vitamins D with Lewis acids[10]. The separation of the two forms of vitamin D and the isomeric isotachysterols is very useful to the analyst because, as it is rare to encounter foods with both forms of vitamin D present, vitamin D_2 and vitamin D_3 can be used as mutual internal standards. This procedure is used in the HPLC method for determination of vitamin D developed at LGC[11]. The stages of this analysis are:

(i) Addition of internal standard to sample, e.g. vitamin D_2 is added if sample contains vitamin D_3.
(ii) Saponification followed by ether extraction.
(iii) Isomerization of vitamins D to the isotachysterols with antimony trichloride.
(iv) Purification of isotachysterols on dry columns of alumina.
(v) HPLC separation of isotachysterols.

Figure 3 shows chromatograms obtained from tinned herrings using the above method. The results are calculated from peak height measurements. This method has also been applied with reasonable success to the analysis of vitamin D in freeze-dried eggs, margarine and fortified infant feeds. This HPLC system has the potential to be used for foods with low levels of vitamin D, e.g. butter (<10 ng/g), but such foods will require more extensive preliminary purification. A method published recently for the HPLC determination of vitamin D in fortified milk[12] used gel permeation, on columns of hydroxyalkoxypropyl Sephadex (HAPS), for preliminary clean-up of extracts prior to HPLC. Recovery experiments performed during the development of this method, however, indicated 10-20% loss of vitamin D. This loss

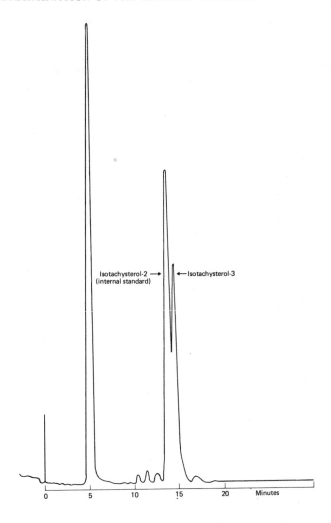

Figure 3 HPLC determination of vitamin D_3 in tinned herring. Column — 5 micron Spherisorb ODS 25 cm × 4 mm; mobile phase — 95% methanol/5% water; detector — UV absorbance 280 nm. (sensitivity 0.01 AUFS)

was eventually attributed to the separation of previtamin D on the HAPS column. Vitamin D_2 or D_3 in solution is in equilibrium with its previtamin, the concentration of which increases with rise in temperature[13]. At 20 °C the equilibrium concentration of previtamin D_2 or D_3 is 7%; this however increases to 20% at 80 °C. The time to reach equilibrium is also temperature-dependent; at 20 °C, 30 days are required to establish equilibrium concentration of previtamin D_2 or D_3 whereas at 80°C only 2.5 h are required. In the above method the previtamin D lost was formed during saponification (30 min at 80°C). The method was therefore modified by saponifying the milk-fat overnight at room temperature to minimize previtamin D formation. The previtamin D content of an extract must be included in any determination of vitamin D content of a food as it represents potential vitamin D activity[14]. Losses of previtamin D during chromatography particularly with HPLC are therefore a problem of which analysts should be aware. Low results for vitamin D content because of previtamin D loss is not a problem with the method developed at LGC because any previtamin D formed during saponification is converted to the isotachysterol along with vitamin D on treatment with antimony trichloride.

An HPLC method has been used to determine the low levels of the metabolite 25-hydroxyvitamin D_3 in bovine tissues[15]. Levels of less than 2 ng/g were determined in bovine muscle using an ultraviolet detector. The tissue extracts did, however, require quite extensive clean-up. The tissues were initially extracted with ethanol/diethyl ether. These extracts were then partitioned between acetonitrile and methylene chloride before chromatography on silica gel. This chromatographic separation was followed by partition chromatography on Celite before application of extracts to HPLC column for final separation. HPLC has also been applied to the analysis of vitamin D, 25-hydroxyvitamin D and 24,25-dihydroxyvitamin D in human plasma[16][17].

It is relatively easy using HPLC to separate vitamin E into its constituent tocopherols and tocotrienols[18]. The difficult separation of β- and γ-tocopherol is quite readily achieved. HPLC has been applied to the analysis of tocopherols in vegetable oils, foods and animal feedstuffs[19-21].

References

1. Bell, J. G. and Christie, A. A. (1974). Gas liquid chromatographic determination of vitamin D in full cream dried milk. *Analyst*, **99**, 385
2. Christie, A. A., Dean, A. C. and Millburn, B. A. (1973). Determination of vitamin E in food by colorimetry and gas liquid chromatography. *Analyst*, **98**, 161
3. Vecchi, M., Vetter, W., Walther, W., Jermstad, S. A. and Schutt, G. W. (1967). Gas-chromatographische und massspektrometrische untersuchung der trimethylsilylather von vitamin A and einiges seiner isomers. *Helv. Chem. Acta*, **50**, 1243
4. Wiggins, R. A. (1976). Chemical analysis of vitamins A, D and E. *Proc. Analyt. Div. Chem. Soc.*, 133
5. Majors, R. E. (1977). Practical considerations in the use of high performance liquid chromatographic columns. *J. Assoc. Offic. Analyt Chemists*, **60**, 186
6. Bell, J. G. (1971). Separation of oil soluble vitamins by partition chromatography on Sephadex LH20. *Chem. Ind.*, 201
7. Wiggins, R. A. and Zai, S. Unpublished results.
8. Egberg, D. C., Heroff, J. C. and Potter, R. H. (1977). Determination of all-trans and 13-*cis* vitamin A in food products by high-pressure liquid chromatography. *J. Agric. Food Chem.*, **25**, 1127
9. Stewart, I. (1977). Provitamin A and carotenoid content of citrus juices. *J. Agric. Food Chem.*, **25**, 1132
10. Wiggins, R. A. (1977). Separation of vitamin D_2 and vitamin D_3 by high pressure liquid chromatography. *Chem. Ind.*, 841

11. Lumley, I., Wiggins, R. A. and Zai, S. Unpublished results.
12. Thompson, J. N., Maxwell, W. B. and L'Abbe, M. (1977). High pressure liquid chromatographic determination of vitamin D in fortified milk. *J. Assoc. Offic. Analyt. Chemists*, **60**, 998
13. Buisman, J. K., Hanewald, K. H., Mulder, F. J., Roborgh, J. R. and Keuning, K. J. (1968). Evaluation of the effect of isomerization on the chemical and biological assay of vitamin D. *J. Pharm. Sci.*, **57**, 1326
14. Mulder, F. J., Devries, E. J. and Borsje, B. (1971). Chemical analysis of vitamin D in concentrates and its problems. XII Analysis of fat-soluble vitamins. *J. Assoc. Offic. Analyt. Chemists*, **54**, 1168
15. Koshy, K. T. and Van Der Slik, A. L. (1977). High-performance liquid chromatographic method for the determination of 25-hydroxycholecalciferol in bovine liver, kidney and muscle. *J. Agric. Food Chem.*, **25**, 1246
16. Jones, G. (1978). Assay of vitamins D_2 and D_3 and 25-hydroxyvitamins D_2 and D_3 in human plasma by high-performance liquid chromatography. *Clin. Chem.*, **24**, 287
17. Lambert, P. W., Syverson, B. J., Arsaud, C. D. and Spelsberg. T. C. (1977). Isolation and quantitation of endogenous vitamin D and its physiologically important metabolites in human plasma by high pressure liquid chromatography. *J. Steroid Biochemistry*, **8**, 929
18. Cavin, J. F. and Inglett, G. E. (1974). High resolution liquid chromatography of vitamin E isomers, *Cereal Chemistry*, **51**, 605
19. Abe, K., Yuguchi, Y. and Katsui, G. (1975). Quantitative determination of tocopherols by high speed liquid chromatography. *J. Nutr. Sci. Vitaminol.*, **21**, 183
20. Soderhjelm, P. and Andersson, B. (1978). Simultaneous determination of vitamins A and E in feeds and foods by reversed phase high-pressure liquid chromatography. *J. Sci. Food Agric.*, **29**, 697
21. Cohen, H. and Lapointe, M. (1978). Method for the extraction and cleanup of animal feed for the determination of liposoluble vitamins A, D and E by high pressure liquid chromatography. *J. Agric. Food Chem.*, **26**, 1210

10
Vitamin A deficiency and excess

G. A. J. PITT

We obtain our vitamin A either as preformed retinol (II) or as carotenoid precursors such as β-carotene (I) which can be split by enzymes in the intestinal wall to give retinol.

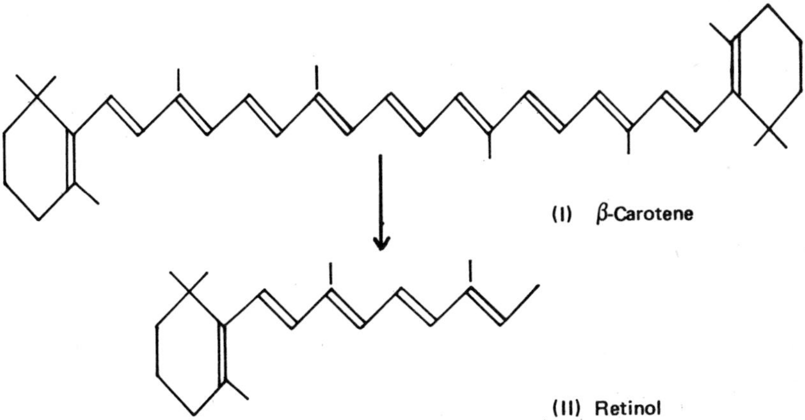

(I) β-Carotene

(II) Retinol

About thirty years ago, it seemed to many that vitamin A deficiency was unlikely to be a very common nutritional problem. Dairy products are a very good source of the vitamin and all green leaves and many yellow vegetables contain carotenoid precursors of vitamin A. Surveys had been carried out in various countries over the previous twenty years on post mortem samples of liver to determine body reserves of vitamin A[1]. The results were interpreted as indicating that most people obtained adequate amounts of the vitamin.

Persuasive evidence came from the 'Sheffield experiment' set up during the Second World War when sixteen human volunteers were given a diet as free as possible from vitamin A[2]. After 18 months, only three had significantly impaired dark adaptation, generally considered to be the most suitable single index of vitamin A deficiency. One subject showed no change in plasma retinol concentration even after 22 months. It was therefore difficult to induce vitamin A deficiency in this country even when every effort was being made to do so. Understandably there tended to be a feeling of complacency, or at least a lack of sense of alarm.

The big swing of opinion began in the 1960s as numerous surveys revealed the vast extent of vitamin A deficiency, particularly in developing countries in the tropical and subtropical regions[3-9]. Contrary to earlier expectations, it has been claimed that after deficiency of total energy and protein, vitamin A is the most common specific dietary deficiency in the world[10] [11].

Many people obtain little preformed retinol because they eat few dairy products; retinol itself has been called the 'prosperity vitamin'[12]. Those eating few dairy foodstuffs are usually dependent on carotene as their source of vitamin A, and many people appear to eat few green vegetables; the lack of dietary carotene is particularly noticeable in populations living on a staple diet of cereals.

Uncomplicated vitamin A deficiency, however, is rarely seen in man. Apart from one freak case in the literature[13], to produce it one has to have human volunteers given a diet complete in all respects except for vitamin A, as was done in the Sheffield experiment. A recent experiment on the same lines has been carried out in the United States on eight males aged between 31 and 43, with the subjects being investigated closely with the more modern aids now available[11] [14].

As always on such diets, the first observable effect was the rapid fall of plasma carotene, an event of no physiological consequence, but reassuring in that it confirmed that the carotene content of the diet was negligible. The plasma retinol concentrations then began to fall slowly, from initial values of $57-78 \mu g/100$ ml, at varying rates in different subjects. In six of the eight subjects the plasma retinol concentration fell to $10 \mu g/100$ ml or below in periods ranging from 359 to 771 days. In one, the plasma retinol concentration had declined to $28-30 \mu g/100$ ml in 361 days, but by then he was showing ophthalmological changes and it was decided to give him vitamin A. In the last subject, plasma retinol concentrations remained around $30 \mu g/100$ ml during 587 days on the deficient diet, and other signs of vitamin A deficiency were minimal.

The plasma retinol concentration fell substantially before any effect was noted on dark adaptation, i.e. the ability to detect very low intensity light. Except for the one subject already mentioned who showed impairment of dark adaptation at plasma retinol concentrations around $30-35 \mu g/100$ ml, the plasma retinol concentration had to fall below $30 \mu g/100$ ml, often below

$10 \mu g/100$ ml and in one subject as low as $3 \mu g/100$ ml, before dark adaptation was affected.

Electroretinograms, which measure the ability of the retina to generate a nerve impulse when light strikes the eye, confirmed the findings on dark adaptation thresholds, but the plasma retinol concentration usually had to fall below $10 \mu g/100$ ml to produce markedly abnormal electroretinograms.

The effects on dark adaptation were rapidly corrected by quite small amounts of vitamin A: $150 \mu g$ retinol/day was adequate; in one subject only $75 \mu g$ retinol/day was needed. To restore normal electroretinograms rather larger amounts — up to $600 \mu g$ retinol/day — were needed, although Sauberlich and co-workers[11] accept that the smaller doses which appeared ineffective might have been adequate had it been ethically acceptable to continue with smaller doses for a longer period.

Faulty dark adaptation was not the first sign of vitamin A deficiency to be detected in their investigation. In almost all cases it was preceded by follicular hyperkeratosis. There has been some controversy in the past about whether follicular hyperkeratosis is a true manifestation of vitamin A deficiency itself[2] or of some concomitant deficiency[15]. The findings in this work[11] indicate that it is a true vitamin A deficiency sign, but the variable experiences of other workers and the subjective element in the assessment of the condition seem to render it a less appropriate method of detecting and monitoring vitamin A deficiency than the more objective measurement of dark adaptation.

The quickest manifestation of any abnormality other than the plasma retinol level was the blood haemoglobin concentration[14]. Despite receiving 18–19 mg iron daily in their diet, the subjects began to manifest a mild degree of anaemia. They were then given therapeutic doses of iron as ferrous gluconate, to supply 310 mg iron daily. This was of little value; there was a transient rise in blood haemoglobin concentration, followed by a fall so long as the plasma retinol concentration remained low, even though 310 mg iron continued to be given. Small doses of vitamin A, however, rapidly restored the blood haemoglobin concentration to normal, even in the absence of therapeutic doses of iron.

The features of vitamin A deficiency in man seen in this experiment were: low plasma retinol concentration[11], low blood haemoglobin concentration[14], follicular hyperkeratosis[11], impairment of dark adaptation[11] and abnormal retinograms[11]. Changes were also mentioned[11] in taste, smell, vestibular function and cerebrospinal fluid pressure, but details of these have not yet been published.

The daily intake of vitamin A recommended by FAO/WHO[16] is $750 \mu g$ retinol. From their work on these volunteers Sauberlich et al.[11] concluded that, although the anaemia was cured with lower doses[14], adult males needed $600 \mu g$ retinol/day to prevent or cure changes in dark vision (as mentioned previously, this might be a slight overestimate) and perhaps a little more to cure the skin lesions.

Normal plasma concentrations of retinol would not have been maintained by 600 μg retinol/day. What minimal concentration of retinol should be maintained in the plasma? Sauberlich and co-workers[11] picked on the figure of 30 μg retinol/100 ml plasma, which does not appear to be overgenerous. Below that, all subjects showed a fall in blood haemoglobin[14]; some showed follicular hyperkeratosis at concentrations slightly above 30 μg retinol/100 ml; one subject even had faulty dark adaptation at that level[11]. Even though 30 μg retinol/100 ml plasma seems a marginal value, this work[11] indicates that it would not be attained in some subjects on a dietary intake of 750 μg retinol/day. There are grounds therefore for believing that the daily allowance of 750 μg retinol recommended by FAO/WHO[16] may be an underestimate.

It should perhaps be remarked in passing that the adoption of 30 μg retinol/100 ml as the minimal acceptable value for plasma vitamin A in children has led to some controversy, and has been considered too high[17].

'Natural' vitamin A deficiency commonly occurs in association with protein-energy malnutrition[3 4 10 19]. It is manifested most obviously and severely in damage to the conjunctiva and cornea in the form of xerophthalmia and keratomalacia[3 4 12]. Xerophthalmia[20] first affects the conjunctiva where epithelial keratinization results in dryness. The cornea is then affected; keratinization of the epithelium and cellular infiltration of the stroma lead to a hazy appearance of the cornea and 'ulceration' occurs. Keratomalacia sets in and the corneal structure melts into a cloudy gelatinous mass; deformation of the eyeball, extrusion of the lens and loss of the vitreous may result. Unless the process is checked early in its development, blindness is irreversible[4 20]. Xerophthalmia, followed by keratomalacia, is an important cause of preventable blindness in young children in many parts of the world[3 5].

Other disease conditions frequently associated with vitamin A deficiency are diarrhoea, measles and many other infections[5]. Although the extent to which these are attributable to vitamin A deficiency is not so clear as with xerophthalmia, there can be little doubt that vitamin A deficiency makes a sizeable contribution to ill-health in many countries. Mortality is high[4 5 6 19]. Those particularly at risk from vitamin A deficiency are children aged between six months and four years.

It is not wholly clear why a diet deficient in protein and energy should so frequently be accompanied by vitamin A deficiency. Besides a shortage of retinol or carotenoids, two extra factors may be involved.

Protein deficiency may result in poor absorption of carotenes and vitamin A from the diet[15 21]. Perhaps the cleavage of carotenoids to retinol is depressed but evidence for this is equivocal[15 22]. Defects in absorption and conversion seem to be improved by the addition not only of protein but also of fat[15 23], and diets resulting in protein-energy malnutrition are usually low in fat.

Vitamin A present as such in the diet, or formed from carotenoids, is carried from the intestine to the liver by the chylomicrons, and it is redistributed from the liver to the tissues that require it on a specific carrier, retinol-binding protein[24][25]. Without plasma retinol-binding protein, retinol will not be properly distributed to the target tissues.

Protein biosynthesis is slowed when dietary protein is short, but at different rates for different proteins. Plasma retinol-binding protein is one of those severely affected[26]. A deficiency of dietary protein by diminishing the amount of plasma retinol-binding protein can inhibit the transport of vitamin A from the liver[18][27-30].

Whatever the reason or reasons — and it may not be the same in all cases — protein-energy malnutrition appears to aggravate vitamin A deficiency[18][19][23]. In countries where vitamin A deficiency is common it is often difficult to know to what extent morbidity stems from lack of vitamin A or from the accompanying protein-energy malnutrition. Wherever vitamin A deficiency or protein-energy malnutrition is treated, it is usually advisable to deal with the other condition at the same time.

Anaemia is also very prevalent in those parts of the world where vitamin A deficiency occurs. Hodges et al.[14] have drawn attention to surveys carried out in eight countries where there was a reasonable amount of iron in the diet, but the vitamin A intake was not so satisfactory: South Vietnam, Chile, northeast Brazil, Uruguay, Ecuador, Venezuela and Guatemala. There was no correlation between iron intake and blood haemoglobin concentration. There was however a statistically significant correlation between plasma retinol concentration and blood haemoglobin concentrations. Hodges et al.[14] mention these correlations (or lack of them) somewhat tentatively, recognizing that no figures were available for plasma folate or B_{12} or for parasitic infestation such as hookworm. Nevertheless, it seems reasonable to assume that vitamin A deficiency will be implicated in many cases of anaemia.

With the extent of vitamin A deficiency in the underdeveloped world established, investigators turned their attention again to the industrialized countries.

It is not easy to determine the vitamin A status of individuals or populations unless they are actually deficient in the vitamin. Plasma retinol concentrations do not correlate well with liver reserves[31]. The accepted way of checking the vitamin A status of a population is to study the vitamin A content of livers taken at post mortem examinations. It was from such studies[1] that it was concluded during the 1930s and 1940s that the vitamin A status of many countries was satisfactory. From time to time other surveys were published which tended to reinforce that impression for countries where one might expect the intake of vitamin A or carotenes to be adequate[32-34]. The publication in 1968–9 of a survey carried out on the liver stores of Canadians was, therefore, an unpleasant surprise[35][36]. The mean value was reasonable – 114 μg retinol/g liver – but many individuals had low reserves, and in 10% of the cases no vitamin A was detected in the liver. There was some geographical variation; in the Montreal area 22% of the subjects had no detectable vitamin

A. Understandably these findings raised some alarm, and similar surveys were made in the United States[11 31 37-39]. They showed that very low liver reserves were much less common in the United States than was found in the Canadian surveys. Some individuals and underprivileged groups were, however, short of vitamin A, and raised a small public health problem.

The Canadian workers[35 36] categorized subjects as having low liver stores if they fell below 40 μg retinol equivalents/g, and this seems to have been accepted by later workers[11 31 37 40]. If the same criterion had been applied, it is likely that some of the earlier surveys would have shown a fair amount of vitamin A deficiency, although at the time they were considered as indicating a satisfactory state of affairs. By arbitrary decisions it is, of course, possible to increase or reduce the declared extent of vitamin A deficiency.

No rechecks had been carried out in this country until very recently when Huque and Truswell[40] reported their results on the vitamin A contents of livers taken at 281 post mortem examinations in London. The mean value was 242 μg retinol/g liver; the median 181 μg/g; and the range 6–1201 μg/g. No individual therefore was found with no reserves; if 40 μg/g was taken as being satisfactory, only 6% of the subjects had less and so would be classified as low and in the risk range. In this country, we appear to be reasonably well provided with vitamin A; indeed of all the surveys carried out, only New Zealand[32] and Ghana[33] have emerged better.

Although they have revealed a few lacunae, these investigations over the last ten years in prosperous countries have tended to confirm the general situation as satisfactory.

In so doing, they bring out strongly the contrast with some (but not all[33 41]) parts of Asia, Africa and Central and South America. Recognizing the seriousness of the situation, governments and agencies have taken steps to deal with vitamin A deficiency. The long term policy is to educate people to eat more green vegetables, but one cannot expect to bring about such dietary changes very quickly[42]. An apparently quicker solution is to give synthetic vitamin A which is available in unlimited amounts at a low price.

How can this be supplied to those that need it, particularly young children? Two basic methods have been tried. One is to fortify with vitamin A a common foodstuff, for example sugar[43-45]. There are practical difficulties in this approach, and India has opted instead for giving massive doses of vitamin A equivalent to 60 000 – 90 000 μg retinol, at intervals of about six months[15]. These are intended to raise the liver stores, which can then be drawn upon to supply the body over a period of months[46-48]. Even if protein nutrition is far from ideal and the production of plasma retinol-binding protein is diminished, enough vitamin A will get to the tissues to prevent vitamin A deficiency, provided liver stores are adequate[15 49]. Such a programme has been used successfully[46 47 50] but others have found it less efficacious, and think that dosing at intervals shorter than six months may be necessary[51-53]. Unfortunately the size of the dose that can be given is limited for fear of inducing hypervitaminosis A[15][47].

It is well known that vitamin A in large doses is toxic — there has been recent correspondence in the *British Medical Journal* on whether Sir Douglas Mawson's expedition to the Antarctic in 1912 came to partial and almost complete disaster because of hypervitaminosis A[54][55]. Despite warnings by professional organizations of the dangers of excessive vitamin A[56][57] a sporadic flow of reports of hypervitaminosis A continues to come through the clinical literature[58][59].

I suspect that this may be partly because people seem curiously reluctant to recognize just how toxic is vitamin A. I have long asserted that considered as a chronic poison vitamin A is probably more harmful than cyanide, but I have usually been disbelieved. The truth of this claim has recently been established by Dr P. N. Okoh[60] in my laboratory.

As a criterion of toxicity we picked on growth, considering that the harmful effect of anything on young animals is likely to be manifested by a slowing down of growth. Twelve male rats (mean weight, 144 g) were divided into three equal groups. Group A received 38.5 μmoles retinyl acetate per rat per day, mixed into the diet to provide a steady supply; Group B received 77 μmoles potassium cyanide per rat per day, mixed into the diet; Group C was the unsupplemented control group. The rats were kept on this diet for three weeks, and their weight changes measured. Group B given potassium cyanide gained 101 g $\pm$ 4 (SEM), essentially the same as the control group C, 100 g $\pm$ 6. Group A given retinyl acetate, gained only 42 g $\pm$ 9, very significantly ($p < 0.005$) less than the control group C, or the cyanide group B. This intake of retinyl acetate had a severely adverse effect on growth, whereas twice the intake of potassium cyanide had none. Retinyl acetate is more than twice as toxic as cyanide in the conditions of this experiment, that is, given mixed into the food over a period of time. (As an acute poison, cyanide is, of course, more effective.)

This is a gimmicky experiment designed to demonstrate the toxicity of vitamin A in an ostentatious way. I think the gimmickry is justifiable to drive home just how toxic is vitamin A. If vitamin A were to be discovered now for the first time, it is highly probable that it would be banned by drug-regulating authorities.

Vitamin A is an essential substance; for lack of it many people in the world suffer ill-health, irreversible blindness or death. But it is a very dangerous substance and it needs to be handled with care.

References

1. Moore, T. (1957). *Vitamin A*, p. 356 (Amsterdam: Elsevier)
2. Hume, E. M. and Krebs, H. A. (1949). Vitamin A requirements of human adults. An experimental study of vitamin A deprivation in man. *Spec. Rep. Ser. Med. Res. Coun.*, **264**

3. McLaren, D. S. (1963). *Malnutrition and The Eye* (New York: Academic Press)
4. Oomen, H. A. P. C. (1961). An outline of xerophthalmia. *Int. Rev. Trop. Med.,* 1, 131
5. Oomen, H. A. P. C., McLaren, D. S. and Escapini, H. (1964). Epidemiology and public health aspects of hypovitaminosis A. A global survey on xerophthalmia. *Trop. Geograph. Med.,* 16, 271
6. McLaren. D. S., Shirajian. E., Tchalian, M. and Khoury, G. (1965). Xerophthalmia in Jordan. *Am. J. Clin. Nutr.,* 17, 117
7. McLaren, D. S. (1966). Present knowledge of the role of vitamin A in health and disease. *Trans. R. Soc. Trop. Med. Hyg.,* 60, 436
8. Chopra, J. G. and Kevany, J. (1970). Hypovitaminosis A in the Americas. *Am. J. Clin. Nutr.,* 23, 231
9. Varela, R. M., Teixeira, S. G. and Batista, M. (1972). Hypovitaminosis A in the sugarcane zone of southern Pernambuco State, Northeast Brazil. *Am. J. Clin. Nutr.,* 25, 800
10. Roels, H. A. (1970). Vitamin A physiology. *J. Am. Med. Assoc.,* 214, 1097
11. Sauberlich, H. E., Hodges, R. E., Wallace, D. L., Kolder, H., Canham, J. E., Hood, J., Raica, N. and Lowry, L. K. (1974). Vitamin A metabolism and requirements in the human studied with the use of labeled retinol. *Vit. Horm.,* 32, 251
12. Oomen, H. A. P. C. (1976). Xerophthalmia. In: Beaton, G. H. and Bengoa, J. M. (eds.). *Nutrition in Preventive Medicine. The Major Deficiency Syndromes, Epidemiology and Approaches to Control,* pp. 94–110. (Geneva: World Health Organization)
13. Sharman, I. M. (1969). An unusual case of self-imposed vitamin A deficiency. *Am. J. Clin. Nutr.,* 22, 1134
14. Hodges, R. E., Sauberlich, H. E., Canham, J. E., Wallace, D. L., Rucker, R. B., Mejca, L. A. and Mohanram, M. (1978). Haemopoietic studies in vitamin A deficiency. *Am. J. Clin. Nutr.,* 31, 876
15. Strikantia, S. G. (1975). Human vitamin A deficiency. *Wld. Rev. Nutr. Diet,* 20, 184
16. Report of a Joint Food and Agriculture Organization/World Health Organization Expert Group. (1967). Requirements of vitamin A, thiamine, riboflavine and niacin. *FAO Nutr. Mrg. Rep. Ser.,* 41
17. Underwood, B. A. (1974). The determination of vitamin A and some aspects of its distribution, mobilization and transport in health and disease. *Wld. Rev. Nutr. Diet,* 19, 123
18. Gopalan, C., Venkatachalam, P. S. and Bhavani, B. (1960). Studies of vitamin A deficiency in children. *Am. J. Clin. Nutr.,* 8, 833
19. Pereira, S. M., Begum, A. and Dumm, M. E. (1966). Vitamin A deficiency in kwashiorkor. *Am. J. Clin. Nutr.,* 19, 182
20. McLaren, D. S., Oomen, H. A. P. C. and Escapini, H. (1966). Ocular manifestations of vitamin A deficiency in man. *Bull. Wld. Hlth. Org.,* 34, 357
absorption of vitamin A palmitate in severe protein malnutrition. *Am. J. Clin. Nutr.,* 7, 185
22. Gronowska-Senger, A. and Wolf, G. (1970). Effect of dietary protein on the enzyme from rat and human intestines which converts β-carotene to retinal. *J. Nutr.,* 100, 300
23. Reddy, V. and Srikantia, S. G. (1966). Serum vitamin A in kwashiorkor. *Am. J. Clin. Nutr.,* 18, 105
24. Kanai, M., Raz, A. and Goodman, D. S. (1968). Retinol-binding protein: the transport protein for vitamin A in human plasma. *J. Clin. Invest.,* 47, 2025
25. Glover, J. (1973). Retinol-binding proteins. *Vit. Horm.,* 31, 1
26. Muhilal, H. and Glover, J. (1974). Effects of dietary deficiencies of protein and retinol on the plasma level of retinol-binding protein in the rat. *Br. J. Nutr.,* 32, 549
27. Arroyave, G., Wilson, D., Méndez, J., Béhar, M. and Scrimshaw, N. S. (1961). Serum and liver vitamin A and lipids in children with severe protein malnutrition. *Am. J. Clin. Nutr.,* 9, 180
28. Arroyave, G., Wilson, D., Contreras, C. and Béhar, M. (1963). Alterations in serum concentration of vitamin A associated with the hypoproteinaemia of severe protein malnutrition. *J. Pediat.,* 62, 920
29. Roels, O. A., Djaeni, S., Trout, M. E., Lauw, T. G., Heath, A., Pocy, S. H., Tarwatjo, M. S. and Suhadi, B. (1963). The effect of protein and fat supplements on vitamin A deficient Indonesian children. *Am. J. Clin. Nutr.,* 12, 380
30. Konno, T., Hansen, J. D. L., Truswell, A. S., Wood-Walker, R. and Becker, D. (1968). Vitamin A deficiency and protein-calorie malnutrition in Cape Town. *S. Afr. Med. J.,* 42, 950

VITAMIN A DEFICIENCY AND EXCESS

31. Underwood, B. A., Siegel, H., Weisell, R. C. and Dolinski, M. (1970). Liver stores of vitamin A in a normal population dying suddenly or rapidly from unnatural causes in New York city. *Am. J. Clin. Nutr.*, **23**, 1037

32. Smith, B. M. and Malthus, E. M. (1962). Vitamin A content of human liver from autopsies in New Zealand. *Br. J. Nutr.*, **16**, 213

33. Dagadu, M. and Gillman, J. (1963). Hypercarotenaemia in Ghanaians. Lancet, **i**, 531

34. Dagadu, J. M. (1967). Distribution of carotene and vitamin A in liver, pancreas and body fat of Ghanaians. *Br. J. Nutr.*, **21**, 453

35. Hoppner, K., Phillips, W. E. J., Murray, T. K. and Campbell, J. S. (1968). Survey of liver vitamin A stores of Canadians. *Can. Med. Ass. J.*, **99**, 983

36. Hoppner, K., Phillips, W. E. J., Erdody, P., Murray, T. K. and Perrin, D. E. (1969). Vitamin A reserves of Canadians. *Can. Med. Assoc. J.*, **101**, 736

37. Raica, N., Scott, J., Lowry, L. and Sauberlich, H. E. (1972). Vitamin A concentration in human tissues collected from five areas in the United States. *Am. J. Clin. Nutr.*, **25**, 291

38. Underwood, B. A. and Denning, C. R. (1972). Blood and liver concentrations of vitamins A and E in children with cystic fibrosis of the pancreas. *Pediatr. Res.*, **6**, 26

39. Mitchell, G. V., Young, M. and Seward, C. R. (1973). Vitamin A and carotene levels of a selected population in metropolitan Washington D.C. *Am. J. Clin. Nutr.*, **26**, 992

40. Huque, T. and Truswell, A. S. (1979). Retinol content of human livers from autopsies in London. *Proc. Nutr. Soc.*, **38** (In press)

41. Suthutvoravoot, S. and Olson, J. A. (1974). Plasma and liver concentrations of vitamin A in a normal population of urban Thai. *Am. J. Clin. Nutr.*, **27**, 883

42. Chen, P. C. Y. (1972). Sociocultural influences on vitamin A deficiency in a rural Malay community. *J. Trop. Med. Hyg.*, **75**, 231

43. Arroyave, G. and Brenes, E. B. (1972). Control de la deficiencia de vitamina A en Guatemala. (Fortificación del azúcar con palmitato de retinol). *Rvta Col. Med. Guatam.*, **23**, 66; *Nutr. Abstr. Rev.* (1973) **43**, 741

44. Arroyave, G., Beghin, I., Flores, M., Soto de Giudo, C. and Ticas, J. M. (1974). Efectos del consumo de azúcar fortificada con retinol, por la madre embarazada y lactante cuya dieta habitual es baja en vitamina A. Estudio de la madre y del niño. *Archivos Latinamericanos de Nutrición*, **24**, 845; *Nutr. Abst. Rev.* (1976), **46**, 58

45. Aranjo, R. L., Souza, M. S. L., Mata-Machado, A. J., Mata-Machado, L. T., Lourdes Mello, M., Costa Cruz, T. A. Vieira, E. C., Souza, D. W. C., Palhares, R. D. and Borges, E. L. (1978). Response of retinol serum levels to the intake of vitamin A-fortified sugar by pre-school children. *Nutr. Rep. Int.*, **17**, 307

46. Srikantia, S. G. and Reddy, V. (1970). Effect of a single massive dose of vitamin A on serum and liver levels of the vitamin. *Am. J. Clin. Nutr.*, **23**, 114

47. Swaminathan, M. C., Susheela, T. P. and Thimmayamma, B. V. S. (1970). Field prophylactic trial with a single annual oral massive dose of vitamin A. *Am. J. Clin. Nutr.*, **23**, 119

48. Olson, J. A. (1972). The prevention of childhood blindness by the administration of massive doses of vitamin A. *Isr. J. Med. Sci.*, **8**, 1199

49. Smith, F. R., Goodman, D. S., Zaklama, M. S., Gabr, M. K., El Maraghy, S. and Patwardhan, V. N. (1973). Serum vitamin A, retinol-binding protein and prealbumin concentrations in protein-calorie malnutrition. 1. A functional defect in hepatic retinol release. *Am. J. Clin. Nutr.*, **26**, 973

50. Susheela, T. P. (1969). Studies on serum vitamin A levels after a single massive oral dose. *Ind. J. Med. Res.*, **57**, 2147

51. Pereira, S. M. and Begum, A. (1969). Prevention of vitamin A deficiency. *Am. J. Clin. Nutr.*, **22**, 858

52. Pereira, S. M. and Begum, A. (1971). Failure of a massive single oral dose of vitamin A to prevent deficiency. *Arch. Dis. Childh.*, **46**, 525

53. Pereira, S. M. and Begum, A. (1973). Retention of a single oral massive dose of vitamin A. *Clin. Sci. Mol. Med.*, **45**, 233

54. Shearman, D. J. C. (1978). Vitamin A and Sir Douglas Mawson. *Br. Med. J.*, **1**, 283

55. Brown, C. T. (1978). Vitamin A and Sir Douglas Mawson. *Br. Med. J.*, **1**, 650

56. American Academy of Pediatrics. Joint Committee Statement: Committees on Drugs and on Nutrition. (1971). The use and abuse of vitamin A. *Pediatrics, (Springfield)*, **48**, 655

57. Canadian Pediatric Society. (1971). The use and abuse of vitamin A. *Can. Med. Assoc. J.*,

104, 521
58. Farrell, G. C., Bhathal, P. S. and Powell, L. W. (1977). Abnormal liver function in chronic hypervitaminosis A. *Am. J. Dig. Dis.*, **22**, 724
59. Shaywitz, B. A., Siegel, N. J. and Pearson, H. A. (1977). Megavitamins for minimal brain dysfunction. A potentially dangerous therapy. *J. Am. Med. Assoc.*, **238**, 1749
60. Okoh, P. N. (1978). Aspects of the metabolism of cyanide in the rat. PhD Thesis: University of Liverpool

11

The importance of sunlight as a source of vitamin D for man

D. E. M. LAWSON

'When you can measure what you are speaking about and express it in numbers, you know something about it, and when you cannot measure it, when you cannot express it in numbers, your knowledge is of a meagre and unsatisfactory kind'.

Since Lord Kelvin expressed this view as the ideal to which scientists should aim there has been an assumption, particularly in modern times, that this aim is achieved. In nutrition, however, there are severe limitations on the extent to which achievement of this aim is possible. In the case of vitamin D nutrition it was over 50 years before even moderate progress, in the form of an assay for vitamin D status, could be made. This long delay before nutritionists had a means of measuring what they were talking about has caused a number of problems. For example there are gaps in our knowledge of vitamin D of which most nutritionists are unaware, because previous generations of investigators either were unable to obtain or did not realize the necessity of obtaining crucial information. Further complications have arisen because of the prolonged period taken to understand the physiology and biochemistry of vitamin D and as a result a number of erroneous attitudes have become established. During this period there were times with very little development in these areas of vitamin D which were taken by some people to mean that a full understanding of these subjects had been reached. In general it is not realized that our views on various aspects related to vitamin D are not on as sound a basis as is believed. In addition it is not appreciated that the recent advances in our knowledge of vitamin D physiology require previous results to be reconsidered. In the last two or three years my colleagues at the Dunn Nutritional Laboratory, Dr D. Fraser and Dr M. Davie, and I have been developing a new approach to nutritional aspects of vitamin D and in this article I would like to describe some results which have caused us to reconsider the need for dietary vitamin D. In the course of this, I will try to illustrate some of the points made above.

Included among the important findings leading to our current understanding of the need of animals for vitamin D and the role of dietary vitamin D in the prevention of rickets are:

(a) The development of an animal model for rickets;
(b) The demonstration of the efficacy of certain foods and of ultraviolet light in curing rickets in animals and later in children;
(c) The isolation and chemical identification of vitamin D_2 and later vitamin D_3;
(d) The use of vitamin D and cod liver oil in the prevention and treatment of rickets;
(e) The fortification of food, including baby foods, with vitamin D.

Thus, some 20 years after the discovery of the existence of vitamin D, rickets had ceased to be a public health problem in Europe and America, apparently as a result of the raising of dietary vitamin D levels. It should be noted, however, that:

1. Vitamin D has never been isolated from any food (chemists used yeast as a source of the provitamin ergosterol for conversion to vitamin D_2);
2. Vitamin D status could, until very recently, only be assessed in a qualitative manner by clinical examination supplemented by some simple biochemical measurements of blood constituents and by radiological examination of the bones;
3. Although the presence of antirachitic activity could be shown in certain foods, particularly those rich in fat, it was never shown that the diet was the source of this activity for man.

Investigation of this last point has had to await the discovery of a chemical method of assessing vitamin D status. Nevertheless, early investigators attempted to draw some valid conclusions on the relative contributions of sunlight and diet in prevention of rickets in man. For example, Chick and her colleagues[1] showed that sunlight was very effective in the cure of rickets in children, yet despite this observation the disease was present in areas of plentiful sunshine such as Southern California[2]. Consequently it appeared that the antirachitic substance in food belonged to that class of substances which animals are no longer able to synthesize (usually because of a deficiency of a certain enzyme but in this case of ultraviolet light) and as a result the compound should be classed as a vitamin. Fortification of the diet with vitamin D seemed appropriate as no food was found which was both rich in this substance and widely consumed. Infants seemed to be particularly prone to develop a deficiency of this vitamin.

A few years ago, however, as a result of major changes in our understanding of the physiology of vitamin D, a chemical method for the assessment of vitamin D was developed. It was found[3] that almost all the antirachitic activity of plasma is due to 25-hydroxy-vitamin D (25-[OH]D) and that vitamin D itself is present only in very small quantities. In addition early studies suggested that relative to other tissues plasma contains a high concentration of antirachitic activity and the highest proportion of the total body activity[4]. An assay for plama 25-[OH]D levels therefore provides a measure of vitamin D status[3], a fortunate conclusion since plasma also contains a specific binding protein for this substance[5]. Using this protein a specific, sensitive and potentially accurate competitive protein binding assay for this metabolite was developed[6]. In contrast an assay for vitamin D, if such a method had been required for assessment of vitamin D status, is only now becoming a practical possibility using high pressure liquid chromatography[7].

We have therefore carried out three studies into factors affecting vitamin D status. In the first study[8] a single sample of blood was obtained from 110 healthy children aged 4–6 years living in Dudley, West Midlands. The samples were obtained over a 17-month period and, not unexpectedly[9] [10], the plasma 25-[OH]D levels showed a marked seasonal variation (Figure 1). The highest values were found in late summer (i.e. August, when mean values were 24.2 ng/ml with a range of 19.7–37.1 ng/ml) and the lowest in late winter (i.e. February, mean value 10.8 ng/ml). No significant difference in plasma 25-[OH]D level was found between boys and girls. This seasonal variation in plasma 25-[OH]D was not unexpected since such seasonal variations have been reported previously[9] [10] and earlier observations[11] on the incidence of rickets suggested that vitamin D status would be variable throughout the year.

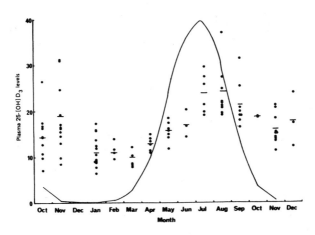

Figure 1 Plasma 25-[OH]D levels (ng/ml) of young children over 15 months from October 1974 to December 1975. Solid line is the variation in the recorded weekly hours of sunshine for this period

A highly significant correlation was found ($p<0.001$) between the rise in plasma 25-[OH]D levels and the hours of sunlight or total ultraviolet light energy during spring and early summer. Interestingly the peak of plasma 25-[OH]D and hours of sunlight are not coincident. Thus, plasma 25-[OH]D began to fall when sunlight was approximately 30 h per week, whereas it had begun to rise in February when there were only about 12 h of sunshine per week. Plasma 25-[OH]D was higher in those children who had a seaside holiday in the year prior to blood sampling than in those children who did not have a holiday away from home that year ($p<0.05$). Serum 25-[OH]D levels did not correlate with dietary intake of vitamin D in the previous 24 h even when allowance was made for seasonal variation in levels. The levels of this steroid were also not affected by intake of eggs by the children nor by the consumption of any pharmaceutical preparations containing vitamin D. While this study emphasizes the role of sunlight in maintaining vitamin D status the assessment of dietary vitamin D by 24 h dietary recall is inadequate to allow any reliable conclusions to be made on the role of dietary vitamin D.

The conclusion from this study therefore is that the rise in 25-[OH]D levels from winter to summer is primarily, if not entirely, due to increased sunlight. The contribution of the diet to serum 25-[OH]D levels cannot be more than the minimum winter level i.e. sufficient to provide 25-[OH]D levels of about 10 ng/ml of plasma. From the relationship of the increase in sunlight hours and serum 25-[OH]D levels it was calculated that the contribution was less than this and possibly sufficient to provide in these children about 7.0 ng/ml of plasma.

Our attention turned next to the vitamin D status of the elderly. The design of this study[12] aimed to take into account some of the deficiencies of the first study; for example, the same group of subjects was studied throughout the period and their dietary habits and outdoor activities were monitored continuously over the same time. A randomly selected group of 23 men and women aged 72–86 years living in Sunderland (latitude 55°) were divided into three groups and each month a group was clinically examined and a blood sample obtained from each subject. Thus in a 12-month period each subject provided four blood samples but the study was continued for 16 months until six samples had been collected. All subjects were in good health for their age. Dietary vitamin D was estimated from daily records of all foods eaten over the whole period supplemented at 6-weekly intervals by one day's quantitative measurement to provide estimates of portion sizes. Food composition tables were used to calculate vitamin D intakes.

The plasma 25-[OH]D levels are shown in Figure 2. A highly significant difference in 25-[OH]D levels ($p<0.001$) was found between the maximum value in the July–August period and the minimum value in the December–May period. The very low plasma 25-[OH]D levels found in this group of people between December and May are of interest. In this period the average value was only 3.6 ng/ml of plasma; nine of the subjects had 25-

[OH]D values of less than 3.0 ng/ml of plasma and in December and January four of the subjects had levels of less than 1.0 ng/ml of plasma. Despite these values there were no clinical signs in these individuals of osteomalacia and plasma Ca, P and alkaline phosphatase levels were all normal. The implications of this have been discussed[13].

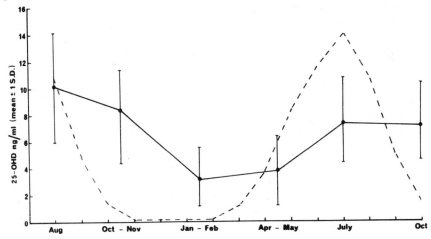

Figure 2 Plasma 25-[OH]D levels of elderly subjects over 16 months from July 1975 to October 1976. The mean and S.D. of the plasma 25-[OH]D values of the 23 subjects recorded during each 3-month period is indicated. Dashed line is the variation in the recorded weekly hours of sunshine for this period

The rise in plasma 25-[OH]D levels was significantly correlated with the increase in the hours of sunshine ($r=0.595$, $p<0.01$) and of ultraviolet light ($r=0.63$, $p<0.001$) occurring throughout this period. The amount of time the subjects spent out of doors was also assessed so that each subject could be allotted an outdoor score. There was little seasonal variation in the pattern of outdoor activities of these people. Again, perhaps not unexpectedly, the 25-[OH]D level in the August–September period and the outdoor score was significantly correlated ($r=0.623$, $p<0.01$) but there was no correlation between the December–April 25-[OH]D level and outdoor score ($r=0.233$). It was notable, however, that a female subject with a relatively high winter 25-[OH]D had one of the highest intakes of dietary vitamin D and yet almost never went out of doors. The omission of this subject from the analysis results in a significant correlation being found between outdoor score and winter 25-[OH]D levels ($r=0.672$, $p<0.01$).

The average daily vitamin D intake over the whole 16-month period was 2.4 μg with a range of 0.7–5.4 μg. There was no seasonal variation. Eggs, fatty fish and butter were the main sources of vitamin D; other than in cakes and pies, margarine was eaten by only three subjects. Ovaltine, which contains added vitamin D, was a good source of the vitamin in three individuals. The vitamin D intake was not correlated with the summer level of 25-[OH]D but there was a significant relationship with the winter plasma 25-[OH]D level ($r=0.55$, $p<0.02$).

The present results clearly show that, in the summer, vitamin D status is primarily governed by the amount of exposure to solar radiation with dietary vitamin D making only a small, almost negligible contribution. In winter, vitamin D is not formed in the skin so the body has to utilize that formed the previous summer and hence solar radiation is still indirectly the primary factor. My colleague, Dr M. Davie, has used the data from the Sunderland subjects to show that the value of the plasma 25-[OH]D reached in summer influences the extent to which the levels fall in winter. He found that subjects with a higher summer level also had a higher winter level ($p>0.05$) and that the relationship was logarithmic[14].

In winter, dietary vitamin D if sufficient, may be important but only in those people who very rarely go outside and who, consequently, have a very low vitamin D status. Dietary vitamin D in any event is insufficient in Britain to prevent the concentration of plasma 25-[OH]D falling in many of the elderly to unsatisfactory levels (i.e. <3.0 ng/ml). The view could be taken that the situation is different in children where the winter levels were appreciably higher than in the elderly. To test the possibility that dietary vitamin D is more effective in younger people in maintaining vitamin D status we carried out a third study[8].

Normal healthy adults in the Cambridge area took 5 μg vitamin D_2 for 28 days from mid-February, 1977. The subjects were chosen because they ate little margarine (the only vitamin D_2-containing food eaten regularly by British adults) and they were asked to avoid eating oily fish and pharmaceutical preparations containing vitamin D for the period of study. Each subject kept a daily record of vitamin D containing foods eaten and vitamin D intake was calculated.

Table 1 Plasma concentrations of 25-[OH]D_2 and 25-[OH]D_3 (ng/ml) in adults before and after taking 5 μg vitamin D_2 daily for 28 days

25-(OH)D_2		25-(OH)D_3	
Day 0	Day 28*	Day 0	Day 28
2.6 ± 1.0	4.0 ± 1.5	9.8 ± 9.2	7.6 ± 4.4

* Significantly higher than corresponding value at day 0 ($p<0.01$)

The results have been published in detail elsewhere but Table 1 shows mean plasma 25-[OH]D_2 and 25-[OH]D_3 levels at the beginning and end of the study. At the commencement of the study mean 25-[OH]D_2 was much lower than mean 25-[OH]D_3. After 5 μg vitamin D_2 daily for one month, there was a small, but significant rise in mean 25-[OH]D_2. Correlation between each individual's intake of vitamin D_3 during the study and serum 25-[OH]D_3 was not significant.

Thus an intake of vitamin D_2 twice the national average intake of this substance leads to a trivial rise in plasma 25-[OH]D_2 of 1.4 ng/ml. This result is entirely consistent with the findings of other investigators on the relationship between the level of oral vitamin D and plasma 25-[OH]D[15-17]. This effect of 5 μg of vitamin D_2 should be contrasted with the effect of ultra-violet light in these subjects as their 25-[OH]D levels in the following September were 22.3 ± 8.5 ng/ml.

Two approaches can be taken to these findings. The traditional view would be that dietary vitamin D intakes of the elderly are inadequate to prevent the fall in winter of their plasma 25-[OH]D to very low levels and consequently dietary vitamin D levels should be raised. This would involve fortification of food to a higher level than that currently practised with margarine. Our results indicate that the intake for the population of this country should be raised to about 5 μg/day in order that the elderly and others with limited exposure to ultraviolet light reach a minimal acceptable vitamin D status i.e. sufficient to prevent occurrence of osteomalacia. Suitable sources of vitamin D eaten by the elderly and which, if eaten very frequently, would provide about 5 μg of vitamin D daily are fatty fish, e.g. sardines, and Ovaltine. However, the consumption of these foods is not widespread and therefore if vitamin D intakes are to be raised to provide protection to groups at risk some other food, more widely eaten than margarine, will have to be fortified.

There is, however, another view of these findings. The vitamin D status of the British population as a whole is undoubtedly adequate for normal health and is maintained as such by the ultraviolet light reaching this country. The contribution by dietary vitamin D is simply unimportant compared to the effects produced by ultraviolet light (plasma 25-[OH]D levels of 20–30 ng/ml). At present the dietary vitamin D equivalent to the amount of ultraviolet light received by the British people is unclear but it is certainly much greater than 5 μg daily, possibly 50 μg/day[16]. If vitamin D is a nutrient then it should be possible to meet the requirements for this substance from a normal diet but in the absence of fortification such levels of vitamin D intake are not only not reached in practice but are extremely unlikely to be reached given the low concentrations at which this substance is usually found. For example, if fortification was not practised, then people who are unable or unwilling to eat fatty fish would need to eat three or four eggs every day of their life simply to produce the very small effect such as that recorded in the results of Table 1. The conclusion I draw from these studies, therefore, is that, physiologically, vitamin D is an essential metabolite with the rare property of being formed by ultraviolet light rather than an enzyme. This statement leads in turn to the question of whether vitamin D is a nutrient. Unfortunately, we cannot at present give a final answer to this point, as none of the studies described here or referred to in this article have considered the situation in infants, perhaps the group most in need of vitamin D, but with the least ability to affect their exposure to sunlight. The source of vitamin D in this group is not obvious since milk, the main and for a time the sole source of food, has such a low level of this substance. However, for children over 2–3 years of age, and for adults, explanations of the development of rickets or osteomalacia must take into account that the sun's ultraviolet radiation is the source of vitamin D. Examples of these conditions developing in older children and adults include rickets in Asian teenagers in Britain, the osteomalacia in young adults[18] in Vienna in 1918–1920 and osteomalacia in the elderly in Britain. Now, of course, it may be argued that the source of vitamin D is an academic point and a low vitamin D status obviously exists and can be counteracted by fortification of food. It is in considering this point that my comments at the beginning of this article are relevant, as questions on the comparative metabolism of vitamin D derived from its two sources have not been investigated in the past, probably because the analytical means of doing so were not available 30–50 years ago when these studies might first have been carried out. The recent report by Fraser[19] indicates that caution should be exercised before assuming that the metabolism of vitamin D from diet and sun is identical. He concluded that 25-hydroxylation of oral vitamin D is not controlled to the same extent as it is of vitamin D released slowly into the circulation and that animals are not able to handle dietary vitamin D in an efficient manner. Such a result is not expected if the body's metabolism was organized to utilize vitamin D in the food. Clinical experience also suggests that dietary- and sunlight-derived vitamin D are not metabolized in the same manner. It has been noted for some time that hypercalcaemia is a frequent complication from the ingestion of large amounts of dietary vitamin D whereas no case of hypercalcaemia has yet been described arising from overexposure to sunlight. Prolonged whole body irradiation cannot raise plasma 25-[OH]D to more than 10% of that which can be achieved by dietary vitamin D. These two observations urge that consideration should be given to the necessity and even wisdom of fortifying food with vitamin D.

If the sun's ultraviolet radiation is the source of vitamin D today then we have to reconsider the cause of rickets before 1940. Changes must have occurred which have reduced the incidence of this disease and no doubt the less polluted atmosphere over our cities had undoubtedly helped. But rickets was also described in places not known for their pollution, e.g. Southern California before 1940 and India, and so other factors have also to be considered, e.g. the improved intake of all nutrients which has occurred in Europe and the USA since 1940 may have a sparing action on the utilization of vitamin D. Alternatively the condition seen today may have a different aetiology from that in the past and may be due to physiological factors, whereas in the past the primary causes were environmental.

Vitamin D deficiency may arise from either a decreased synthesis of vitamin D in the skin or an increased catabolism of the vitamin and its metabolites.

Recent results from Dr Davie and myself show that the elderly and probably coloured teenagers are able to synthesize vitamin D as well as young white people[20]. It may be, therefore, that rickets in Asians or the low vitamin D status of the elderly are due to increased rates of metabolism (or catabolism). The results in Figures 1 and 2 show that vitamin D metabolism occurs at a rapid rate but again there is little information on this aspect of vitamin D to be found in the older literature. The only study carried out so far[21] has indicated the importance adipose tissue may play in maintaining reserves of vitamin D during periods of deprivation.

CONCLUSIONS

In this article I have described the variations which take place in vitamin D status of British people through a year. In the first part I made some cautionary points against assuming that our knowledge of vitamin D physiology is on a sound basis. Results are presented showing that irrespective of the situation in the past, vitamin D status at present is controlled both in summer and winter by exposure to ultraviolet light. Dietary vitamin D is at too low a level for this source to have any significant effect on vitamin D status. These observations have come about because we can now fulfil Lord Kelvin's dictum quoted earlier. However, now that the levels of vitamin D and its metabolites in tissues can be measured other essential experiments can be carried out. In particular information on rates of change of vitamin D and its metabolites in tissues under various conditions should be sought and perhaps a full understanding of the aetiology of rickets may be possible.

References

1. Chick, H., Dalyell, E. J., Hume, E. M., McKay, H. M. M., Smith, H. H. and Wimberger, H. (1922). Observations upon the prophylaxis and cure of rickets at the University Kinderklinik, Vienna. *Lancet*, **11**, 7
2. Moore, C. V., Brodie, J. L., Thornton, A., Lessem, A. M. and Cordua, O. B. (1937). Failure of abundant sunshine to protect against rickets. *Am. J. Dis. Child*, **54**, 1227
3. Edelstein, S., Charman, M., Lawson, D. E. M. and Kodicek, E. (1974). Competitive protein-binding assay for 25-hydroxycholecalciferol. *Clin. Sci. Mol. Med.*, **46**, 231
4. Quarterman, J., Dalgarno, A. C. and Adam, A. (1964). Some factors affecting the level of vitamin D in the blood of sheep. *Br. J. Nutr.*, **18**, 79
5. Haddad, J. G. and Chyu, K. J. (1971). 25-hydroxycholecalciferol binding globulin in human plasma. *Biochim. Biophys. Acta*, **248**, 471
6. Lawson, D. E. M. (1979). In *Methods in Enzymology*, D. B. McCormick and L. D. Wright (eds.), (New York: Academic Press). (In press)
7. Jones, G. (1978). Assay of vitamins D_2 and D_3 and 25-hydroxyvitamins D_2 and D_3 in human plasma by high performance liquid chromatography. *Clin. Chem.*, **24**, 287
8. Poskitt, E. M., Cole, T. and Lawson, D. E. M. (1979). Diet, sunlight and 25-hydroxy vitamin D in healthy children and adults. *Br. Med. J.*, **1**, 221
9. Stamp, T. C. B. and Round, J. M. (1974). Seasonal changes in human plasma levels of 25-hydroxyvitamin D. *Nature*, **247**, 563
10. McLoughlin, M. Raggatt, P. R., Fairney, A., Brown, D. J., Lester, E. and Wills, M. R. (1974). Seasonal variations in serum 25-hydroxycholecalciferol in healthy people. *Lancet*, **i**, 536
11. Wimberger, H. In *Studies of Rickets in Vienna 1919–1922*. MRC Special Report Series No. 77, p. 121. (London: HMSO)

12. Lawson, D. E. M., Paul, A. A., Black, A. E., Cole, T. J., Mandal, A. P. and Davie, M. (1979). (In preparation)
13. Davie, M., Lawson, D. E. M. and Jung, R. T. (1978). Low plasma 25-hydroxyvitamin D without osteomalacia. *Lancet*, **i**, 820
14. Lawson, D. E. M. and Davie, M. Aspects of the metabolism and function of vitamin D. *Vitamin Horm.*, (In press)
15. Stamp, T. C. B. (1975). Factors in human vitamin D nutrition and in the production and cure of classical rickets. *Proc. Nutr. Soc.*, **34**, 119
16. Stamp, T. C. B., Haddad, J. G. and Twigg, C. A. (1977). Comparison of oral 25-hydroxy-cholecalciferol, vitamin D and ultraviolet light as determinants of circulating 25-hydroxy-vitamin D. *Lancet*, **i**, 1341
17. Somerville, P. J., Lien, J. W. K. and Kaye, M. J. (1977). The calcium and vitamin D status in an elderly female population and their response to administered supplemental vitamin D_3. *Geronto.*, **32**, 659
18. Hume, E. M. and Nirenstein, J. (1921). Comparative treatment of cases of hunger – osteomalacia in Vienna, 1920 as out patients with cod liver oil and plant oil. *Lancet*, **ii**, 849
19. Fraser, D. R. The economy of vitamin D metabolism. In: *Pediatric Diseases Related to Calcium*, C. Anast and H. F. Delvea (eds.). (New York: Elsevier)
20. Davie, M. and Lawson, D. E. M. Quantitation of plasma 25-hydroxyvitamin D response to ultraviolet irradiation over a controlled area in young and elderly subjects. (In preparation)
21. Rosenstreich, S. J., Rich, C. and Volwiler, W. (1971). Deposition in and release of vitamin D_3 from body fat: evidence for a storage site in the rat. *J. Clin. Invest.*, **50**, 679

12
Vitamin E in human nutrition

M. S. LOSOWSKY

INTRODUCTION

Vitamin E was discovered over fifty years ago by the demonstration that pregnant rats fed rancid lard diets containing all of the then known nutritional factors did not go on to term. Resorption of the fœtus was prevented by lettuce and wheat germ oil, thus implying the presence of an additional nutritional factor which was rapidly shown to be lipid soluble. In 1936 an active substance was isolated and given the name tocopherol[1].

SUBSTANCES WITH VITAMIN E ACTIVITY

It is now clear that there are several different biologically active substances. The natural substances fall into two series, the tocopherols and the tocotrienols. In addition there are some synthetic amino and methyl amino compounds with vitamin E activity.

The biological importance of these substances depends not only on their activity in the body but also on their availability in food and on the efficiency of their absorption. Alpha-tocopherol is the most active of all the substances and is also biologically the most important. Apart from α-tocopherol, the only others of biological importance are β-tocopherol, γ-tocopherol and α-tocotrienol. Although there are varying estimates, it is usually suggested that β-tocopherol is about 30% as active and -tocopherol about 10% as active as α, with α-tocotrienol suggested as perhaps 40% as active as α-tocopherol. Alpha-tocopherol is usually taken to represent about 80% of the vitamin E activity of the diet and for nutritional purposes this is sufficiently accurate in most circumstances[2].

DEFICIENCY IN ANIMALS

There is a very large literature demonstrating a wide range of severe defects due to vitamin E deficiency in many species of animals. These include the rat, chick, dog, mink, pig, horse, cow, mouse, cat, sheep, monkey and many others; the disorders include abortion, infertility, muscular dystrophy, exudative diathesis (oedema), hepatic necrosis, anaemia and many others.

The particular syndrome which occurs with vitamin E deficiency varies with the species but also with a variety of other factors including the dietary intake of selenium, the dietary intake of polyunsaturated fatty acids and the dietary intake of sulphur-containing amino acids[3][4]; various synthetic anti-oxidants can also modify the syndrome.

In view of the widespread effects in so many species of animals, it has been suggested that it is intrinsically likely that deficiency in man is of some importance.

INTAKE AND ABSORPTION OF VITAMIN E

The major source of vitamin E in the diet of man is vegetable oils and margarine. Appreciable amounts of the vitamin do, however, occur in a wide range of foodstuffs including milk products, eggs, cereals, meat, fish and green vegetables.

On average about 30% of the tocopherols present in the diet are absorbed in to the body[5]. The percentage absorption is less for pharmacological doses and greater for tracer doses[5]. Thus, experimentally, absorption varies inversely with the intake but this is probably not of importance in the relatively limited range of intakes seen in normal diets. Absorption seems to be unaffected by the presence of vitamin E deficiency in the body[5]; thus there is no compensatory increase in absorption in the presence of deficiency as occurs, for example, with iron deficiency.

Tocopherols, being fat soluble, are absorbed along with dietary fat; bile and pancreatic juice are needed for efficient absorption. Absorption varies with the efficiency of fat absorption[5] and anything which causes malabsorption of fat also causes malabsorption of tocopherol[7][8]. Although information is not available, it seems likely that not all substances with vitamin E activity are equally well absorbed[9].

DETECTION OF VITAMIN E DEFICIENCY IN MAN

The simplest and most direct evidence of vitamin E deficiency is obtained by the measurement of the plasma tocopherol level. It has been well shown, for example, that low plasma tocopherol levels are found consistently in patients with severe malabsorption, cystic fibrosis being a good example[10][11]. There is now ample evidence of the normal values for plasma α-tocopherol concentration and ranges for different ages in both sexes have been reported[12][13].

However, the plasma tocopherol level is not necessarily very meaningful when taken in isolation. Plasma tocopherol is carried in the lipoproteins, and in patients who are not vitamin E deficient plasma tocopherol correlates very well with various lipoprotein fractions, the best correlation being obtained with the β-lipoprotein and an even better correlation with the total plasma lipid level[14-16]. Thus, the plasma tocopherol level varies not only with the tocopherol status of the individual but also with the plasma lipid level. Therefore a more meaningful way of expressing the plasma tocopherol is as a ratio to the total plasma lipid[15]. Representative lower limits of normal for plasma tocopherol might be taken as 5 μg/ml or as 0.8 mg/g lipid. It should be emphasized, however, that these limits vary with the methods and from one laboratory to another.

Another widely used indicator of vitamin E status is the extent of the *in vitro* haemolysis of red cells when incubated with hydrogen peroxide. A high degree of peroxide haemolysis accompanies vitamin E deficiency. When performed very carefully, under standard conditions, the test is a good indicator of vitamin E status but it must be emphasized that results vary with minor differences in the method, meticulous attention to detail is necessary and each laboratory needs to standardize its own test and establish its own normal range. The curve relating plasma tocopherol to peroxide haemolysis is probably sigmoid in form: when the plasma tocopherol level falls below a certain value the increase in haemolysis is rapid and there comes a stage where further fall does not result in a greater percentage haemolysis[11 17]. It is not clear whether peroxide haemolysis correlates with the total plasma tocopherol level only, or whether there is a relationship to the plasma lipids. Unfortunately, the finding of a high peroxide haemolysis value is not specific for vitamin E deficiency[18], the test must therefore be interpreted in relation to other information.

Other findings in vitamin E deficiency in man are a high level of urinary creatine and a reduction in the survival of red blood cells. Of course, neither of these tests is specific.

All of the manifestations of vitamin E deficiency described above are reversible by the administration of vitamin E. The only irreversible indicator of vitamin E deficiency in man is the finding of ceroid pigment in tissues[8 19]. As far as is known this does not result in functional impairment and there are no associated symptoms or signs purely due to the ceroid.

CLINICAL DEFICIENCY OF VITAMIN E

Good evidence of deficiency of vitamin E has been described in a number of clinical situations. These include infancy (especially pre-term), small bowel disease[20-22], pancreatic disease[11], gastric surgery[23], alcoholism[20], liver disease[22 24] and obstructive jaundice[24].

The most important clinical situation in which vitamin E deficiency occurs is in infants[25], especially pre-term infants, in whom it leads to a haemolytic anaemia which is associated with thrombocytosis, generalized irritability and oedema[26] and is exacerbated by early therapy with iron[27]. In such infants vitamin E deficiency may lead to a greater susceptibility to retro-lental fibro-plasia[28 29] and bronchopulmonary dysplasia[29 30], especially following the use of oxygen therapy. Several factors contribute to the occurrence of deficiency in infants[31]. In pre-term infants there is a need for rapid growth and presumably for a plentiful supply of vitamin E to the tissues and there may be impaired absorption due to immaturity of the digestive organs[32]. Such infants are often in need of oxygen therapy and the suggested antioxidant action of vitamin E may be relevant to the undue susceptibility to complications of oxygen therapy. There is poor placental transfer of vitamin E[33] so that, even if the mother is not deficient, vitamin E nutrition in the infant is precarious[34]. Transfer in the milk is efficient[33] but the content of vitamin E in cows' milk is less than in human milk. Some of the artificial formula diets used in pre-term and small infants may be poor in vitamin E and some contain poly-unsaturated fatty acids which increase the requirement for vitamin E[26 35]. Finally, even though such infants may be iron deficient, under certain circumstances the administration of iron supplements may exacerbate the anaemia caused by the vitamin E deficiency[36].

There is no clinical syndrome associated with vitamin E deficiency in adult man. The diminution in red cell survival is not severe enough to lead to anaemia and can be easily reversed by supplementary tocopherol[11 37]. The creatinuria[20 38] presumably reflects muscle damage[8] but associated symptoms and signs seem not to occur.

DIETARY REQUIREMENT FOR VITAMIN E

It is not possible to state a single dietary requirement for vitamin E since the situation is complicated by a number of factors[39-41]. There is a clear, direct relationship between the dietary intake of polyunsaturated fatty acids and the dietary requirement for vitamin E; there is an equally clear direct relationship between the tissue content of polyunsaturated fatty acids and the dietary requirement for vitamin E[42]. It is fortunate that most foods which contain large amounts of polyunsaturated fatty acids also contain large amounts of vitamin E. Thus, in subjects on a steady intake of the usual foods containing polyunsaturated fatty acids, it is likely that the intake of vitamin E will be commensurate and deficiency will not develop. Theoretically, deficiency could, however, be induced by those foods which contain polyunsaturated fatty acids but do not contain a correspondingly high amount of vitamin E. Furthermore, after prolonged intake of polyunsaturated fatty acids the tissue content is high and a change to a saturated fat diet and thus a lower vitamin E intake might lead to deficiency.

Other factors which are known to affect the requirement for vitamin E in the diet include the dietary contents of selenium, sulphur-containing amino acids and vitamin A. These factors are well established in animals and may be relevant to man but little clinical information is available. A further com-plicating factor in the assessment of the dietary requirement for vitamin E is the very great variation in the vitamin E contents of raw materials, foods and diets[2 43]. The vitamin E content of individual raw materials such as vegetables is affected by processing, storage, freezing and the season of the year[43]. In the preparation of foodstuffs, probably the manoeuvre which most reduces the vitamin E intake of foods is deep frying.

Indirect assessments of dietary intake of vitamin E from food tables are unreliable and exaggerated estimates of daily intakes have been quoted. Measurements of vitamin E intake of whole daily diets as eaten in the United Kingdom show surprisingly low values[2]. The range is of the order of 1–8 mg α-tocopherol. In the United States the values are rather higher, probably largely due to the higher intake of polyunsaturated fatty acids, the values there averaging 3–18 international units or so per day[6][43] (one international unit is the activity of 1 mg DL-α-tocopherol), γ-tocopherol forming substantial, even the major, constituent of the vitamin E in the diet[44].

It thus seems that the minimum requirement for vitamin E is relatively small and this combined with the large variety of foods which contain significant amounts of the vitamin, the fact that man eats a varied diet, and the large stores in the body, perhaps averaging several grams, helps to explain the rarity of deficiency in man, except in the neonate or in association with gross fat malabsorption or severe undernutrition.

In the United Kingdom a specific dietary requirement is not stated but about 10 mg/d is suggested as being the usual dietary intake[45]. In the USA the National Academy of Sciences in 1974 suggested that a range of 10–20 international units would be present in diets supplying 1800–3000 kcal[46]. It seems likely that these quoted figures are reasonable and offer a margin of safety but it must be remembered that no single figure is applicable to all circumstances.

PHARMACOLOGICAL USES FOR VITAMIN E

There are many disorders for which large doses of vitamin E have been claimed to be of benefit or detriment. These include ischaemic heart disease[47-50], ageing[51], thrombophlebitis[52], infertility, malignancy, and protection against toxicity from various substances[53]. The evidence for benefit in these is, at best, poor with the exception of some evidence of protection by vitamin E against certain toxins such as paracetamol[54], air pollutants and carbon tetrachloride.

In practice, vitamin E is not recommended for any of these conditions or for very many others for which its use has been advocated. There is, however, a case to be made for vitamin E therapy in intermittent claudication[55]. Some trials suggest benefit for patients treated with large doses for long periods[56-59]. Vitamin E therapy in this condition has received a certain amount of acceptance although not all trials have shown positive results and there remains considerable doubt.

One other condition in which there is suggestive but not conclusive evidence of a benefit from vitamin E in high doses is the disease of a-β-lipoproteinaemia. This is an extremely rare disorder in which there is an absence of β-lipoproteins in the serum and associated with this are intestinal malabsorption with steatorrhoea, retinitis pigmentosa, spiky misshapen red cells with a diminution in red cell survival, and a progressive neurological deficit. Vitamin E therapy can certainly improve the survival of the red cells *in vitro* but does not alter the fundamental abnormality of their shape, and it is at least possible that vitamin E may reduce the rate of deterioration of the neurological syndrome[60] but proof of this is unlikely to be achieved.

Another circumstance in which vitamin E has been recommended in large doses is to improve function and thus athletic performance. There is now a consensus, based on a number of careful studies, that vitamin E is of no benefit in this situation[61] [62].

RECENT WORK ON VITAMIN E

There continues to be an outpouring of papers suggesting new uses for vitamin E. Vitamin E deficiency has been suggested as a possible cause of sudden death in infants[63] but this seems unlikely[64] [65]. Abnormalities of platelet function[66], platelet ulatrastructure[67], platelet aggregation[68-71], and platelet prostaglandin synthesis during blood coagulation[72] [73] have been demonstrated in the presence of vitamin E deficiency. No clinical concomitants of these findings have been described and there is no good evidence of clinical benefit from vitamin E therapy in relation to these findings.

Vitamin E deficiency has been described in association with Caffey's disease[75] and thalassaemia[76] [77] but the significance of these findings is not clear. An effect of vitamin E deficiency on drug hydroxylating enzymes has been described[78] and might conceivably be of clinical significance although this has not been demonstrated.

Vitamin E excess has been suggested as being a contributory factor in malignant hyperthermia during anaesthesia but this is entirely speculative[74].

It is possible that α-tocopherol pretreatment may increase the bone marrow toxicity of certain antitumour drugs[79].

TOXICITY

Very large doses of tocopherol in animals have been shown to produce toxic effects. These include depression of growth, decrease in thyroid and gonadal function, increase in the requirements for vitamin E and vitamin K, and liver damage.

In man, relatively large doses, of the order of several hundred milligrams per day, have been given in careful studies and in general no toxicity has been shown[80]. There has, however, been a note of caution sounded on occasions and it seems that, with doses greater than several hundred milligrams per day, individual subjects may show some toxic effects. Those reported include gastrointestinal symptoms, angular stomatitis, lethargy, muscular weakness, creatinuria, increased serum creatine phosphokinase, thrombophlebitis, interference with the action of vitamin K in a patient taking oral anti-coagulants and interference with the action of iron in relieving iron deficiency anaemia[81-83]. These have all been rapidly reversed on stopping the vitamin and no serious sequelae appear to have resulted.

CONCLUSIONS

The place of vitamin E in human nutrition remains uncertain except in the infant. It seems likely that, at all ages, there is a requirement for some vitamin E in the diet for optimum health. The amount of this requirement depends on a number of other factors.

Vitamin E deficiency occurs in the infant, especially if pre-term, and the consequences are important to recognize and treat. In adult man, the requirement for vitamin E is small and its availability widespread and thus deficiency occurs only in association with certain diseases, chiefly those causing fat malabsorption; the consequences of the deficiency are apparently minor, in contrast to most other species.

Vitamin E in large doses may be of benefit in intermittent claudication and a–β-lipoproteinaemia. Claims for benefit in other disorders are not well founded. The toxicity of large doses of vitamin E is small.

Intriguing effects of vitamin E are still being described. These may further our understanding of fundamental aspects of tissue structure and function and perhaps lead to new therapeutic uses for vitamin E.

References

1. Evans, H. M., Emerson, O. H. and Emerson, G. A. (1936). The isolation from wheat germ oil of an alcohol, α-tocopherol, having the properties of vitamin E. *J. Biol. Chem.*, **113**, 319
2. Smith, C. L., Kelleher, J., Losowsky, M. S. and Morrish, N. (1971). The content of vitamin E in British diets. *Br. J. Nutr.*, **26**, 89
3. Scott, M. L. (1970). Nutritional and metabolic interrelationships involving vitamin E, selenium and cystine in the chicken. *Int. J. Vit. Res.*, **40**, 334
4. Wertz, J. M., Seward, C. R., Hove, E. L. and Adkins, J. S. (1968). Interrelationship of dietary glycine, methionine and vitamin E in the rat. *J. Nutr.*, **94**, 129
5. Losowsky, M. S., Kelleher, J., Walker, B. E., Davies, T. and Smith, C. L. (1972). Intake and absorption of tocopherol. *Ann. N.Y. Acad. Sci.*, **203**, 212
6. Bieri, J. G. and Evarts, R. P. (1973). Tocopherols and fatty acids in American diets. *J. Am. Diet. Ass.*, **62**, 147
7. Kelleher, J. and Losowsky, M. S. (1970). The absorption of α-tocopherol in man. *Br. J. Nutr.*, **24**, 1033
8. Binder, H. J., Herting, D. C., Hurst, V., Finch, S. C. and Spiro, H. M. (1965). Tocopherol deficiency in man. *N. Eng. J. Med.*, **273**, 1289
9. Blomstrand, R. and Forsgren, L. (1968). Labelled tocopherols in man. *Int. J. Vit. Nutr. Res.*, **38**, 328
10. Underwood, B. A., Denning, C. R. and Navab, M. (1972). Polyunsaturated fatty acids and tocopherol levels in patients with cystic fibrosis. *Ann N.Y. Acad. Sci.*, **203**, 237
11. Farrell, P. M., Bieri, J. G., Fratantoni, J. F., Wood, R. E. and Di Sant' Agnese, P. A. (1977). The occurrence and effects of human vitamin E deficiency. *J. Clin. Invest.*, **60**, 233
12. Kelleher, J. and Losowsky, M. S. (1978). Vitamin E in the elderly. In: *Tocopherol, Oxygen*

and Biomembranes. C. de Duve and O. Hayaishi (eds.). pp. 311–327. (Elsevier/ North Holland Biomedical Press)

13. Chen, L. H., Hsu, S. J., Huang, P. C. and Chen, J. S. (1977). Vitamin E status of Chinese population in Taiwan. *Am. J. Clin. Nutr.*, **30**, 728

14. Davies, T., Kelleher, J. and Losowsky, M. S. (1969). Interrelation of serum lipoprotein and tocopherol levels. *Clin. Chim. Acta.*, **24**, 431

15. Horwitt, M. K., Harvey, C. C., Dahm, C. H. Jr. and Searcy, M. T. (1972). Relationship between tocopherol and serum lipid levels for determination of nutritional adequacy. *Ann. N.Y. Acad. Sci.*, **203**, 223

16. Kayden, H. J. (1978). The transport and distribution of α-tocopherol in serum lipoproteins and the formed elements of the blood. In: *Tocopherol, Oxygen and Biomembranes.* C. de Duve and O. Hayaishi (eds.). pp. 131–142. (Elsevier/ North Holland Biomedical Press)

17. Leonard, P. J. and Losowsky, M. S. (1967). Relationship between plasma vitamin E level and peroxide hemolysis test in human subjects. *Am. J. Clin. Nutr.*, **20**, 795

18. Nitowsky, H. M. and Tildon, J. T. (1956). Some studies of tocopherol deficiency in infants and children. III. Relation of blood catalase activity and other factors to hemolysis of erythrocytes in hydrogen peroxide. *Am. J. Clin. Nutr.*, **4**, 397

19. Bauman, M. B., DiMase, J. D., Oski, F. and Senior, J. R. (1968). Brown bowel and skeletal myopathy associated with vitamin E depletion in pancreatic insufficiency. *Gastroenterology*, **54**, 93

20. Losowsky, M. S. and Leonard, P. J. (1967). Evidence of vitamin E deficiency in patients with malabsorption or alcoholism and the effects of therapy. *Gut*, **8**, 539

21. McWhirter, W. R. (1975). Plasma tocopherol in infants and children. *Acta Paediatr. Scand.*, **64**, 446

22. Göransson, G., Nordén, Å. and Åkesson, B. (1973). Low plasma tocopherol levels in patients with gastrointestinal disorders. *Scan. J. Gastroenterol.*, **8**, 21

23. Leonard, P. J., Losowsky, M. S. and Pulvertaft, C. N. (1966). Vitamin E levels after gastric surgery. *Gut*, **7**, 578

24. Powell, L. W. (1973). Vitamin E deficiency in human liver disease and its relation to haemolysis. *Austr. N.Z. J. Med.*, **3**, 355

25. Jansson, L., Holmberg, L., Nilsson, B. and Johansson, B. (1978). Vitamin E requirements of preterm infants. *Acta Paediatr. Scand.*, **67**, 459

26. Ritchie, J. H., Fish, M. B., McMasters, V. and Grossman, M. (1968). Edema and hemolytic anemia in premature infants. *N. Engl. J. Med.*, **279**, 1185

27. Williams, M. L., Shott, R. J., O'Neal, P. L. and Oski, F. A. (1975). Role of dietary iron and fat on vitamin E deficiency anemia of infancy. *N. Engl. J. Med.*, **292**, 887

28. Johnson, L., Schaffer, D. and Boggs, T. R. Jr. (1974). The premature infant, vitamin E deficiency and retrolental fibroplasia. *Am. J. Clin. Nutr.*, **27**, 1158

29. Oski, F. A. (1977). Metabolism and physiologic roles of vitamin E. *Hospital Practice*, 79

30. Ehrenkranz, R. A., Bonta, B. W., Ablow, R. C. and Warshaw, J. B. (1978). Amelioration of bronchopulmonary dysplasia after vitamin E administration. *N. Engl. J. Med.*, **299**, 564

31. Leading Article. (1977). Vitamin E for babies. *Lancet*, **ii**, 1268

32. Melhorn, D. K. and Gross, S. (1971). Vitamin E-dependent anemia in the premature infant. II. Relationships between gestational age and absorption of vitamin E. *J. Pediatr.*, **79**, 581

33. Martin, M. M. and Hurley, L. S. (1977). Effect of large amounts of vitamin E during pregnancy and lactation. *Am. J. Clin. Nutr.*, **30**, 1629

34. Mino, M., Nishino, H., Yamaguchi, T. and Hayashi, M. (1977). Tocopherol level in human fetal and infant liver. *J. Nutr. Sci. Vitaminol.*, **23**, 63

35. Hassan, H., Hashim, S. A., Van Itallie, T. B. and Sebrell, W. H. (1966). Syndrome in premature infants associated with low plasma vitamin E levels and high polyunsaturated fatty acid diet. *Am. J. Clin. Nutr.*, **19**, 147

36. Melhorn, D. K. and Gross, S. (1971). Vitamin E-dependent anemia in the premature infant. I. Effects of large doses of medicinal iron. *J. Pediatr.*, **79**, 569

37. Leonard, P. J. and Losowsky, M. S. (1971). Effect of α-tocopherol administration on red cell survival in vitamin E-deficient human subjects. *Am. J. Clin. Nutr.*, **24**, 388

38. Nitowsky, H. M., Tildon, J. T., Levin, S. and Gordon, H. H. (1962). Studies of tocopherol deficiency in infants and children. VII. The effect of tocopherol on urinary, plasma and muscle creatine. *Am. J. Clin. Nutr.*, **10**, 368

39. Horwitt, M. K. (1974). Status of human requirements for vitamin E. *Am. J. Clin. Nutr.*, **27**, 1182
40. Bieri, J. G. (1975). Vitamin E. *Nutr. Rev.*, **33**, 161
41. Review. (1973). Selenium: An essential element for glutathione peroxidase activity. *Nutr. Rev.*, **31**, 289
42. Witting, L. A. (1974). Vitamin E-polyunsaturated lipid relationship in diet and tissues. *Am. J. Clin. Nutr.*, **27**, 952
43. Bunnell, R. H., Keating, J., Quaresimo, A. and Parman, G. K. (1965). Alpha-tocopherol content of foods. *Am. J. Clin. Nutr.*, **17**, 1
44. Bieri, J. G. and Evarts, R. P. (1974). Gamma-tocopherol: metabolism, biological activity and significance in human vitamin E nutrition. *Am. J. Clin. Nutr.*, **27**, 980
45. Department of Health and Social Security. (1969). Recommended intakes of nutrients for the United Kingdom. *Reports on Public Health and Medical Subjects*, No. 120. (London: HMSO)
46. *Recommended Dietary Allowances*, 8th Edition. (1974). (Washington D.C.: National Academy of Sciences)
47. Hodges, R. E. (1973). Vitamin E and coronary heart disease. *J. Am. Diet. Ass.*, **62**, 638
48. Olson, R. E. (1973). Vitamin E and its relation to heart disease. *Circulation*, **48**, 179
49. Andersen, T. W. and Reid, D. B. W. (1974). A double-blind trial of vitamin E in angina pectoris. *Am. J. Clin. Nutr.*, **27**, 1174
50. Toone, W. M. and Cohen, H. M. (1973). Letters. *N. Engl. J. Med.*, **289**, 979
51. Leading Article. (1968). Slower ageing. *Lancet*, **ii**, 1281
52. Roberts, H. J. (1978). Letter. *Lancet*, **i**, 49
53. Review. (1977). Vitamin E. *Nutr. Rev.*, **35**, 57
54. Walker, B. E., Kelleher, J., Dixon, M. F. and Losowsky, M. S. (1974). Vitamin E protection of the liver from paracetamol in the rat. *Clin. Sci. Mol. Med.*, **47**, 449
55. Marks, J. (1962). Critical appraisal of the therapeutic value of α-tocopherol. *Vit. Horm.*, **20**, 573
56. Livingstone, P. D. and Jones, C. (1958). Treatment of intermittent claudication with vitamin E. *Lancet*, **ii**, 602
57. Williams, H. T. G., Clein, L. J. and Macbeth, R. A. (1962). Alpha-tocopherol in the treatment of intermittent claudication: A preliminary report. *Can. Med. Ass. J.*, **87**, 538
58. Boyd, A. M. and Marks, J. (1963). Treatment of intermittent claudication: A reappraisal of the value of α-tocopherol. *Angiology*, **14**, 198
59. Haeger, K. (1974). Long-time treatment in intermittent claudication with vitamin E. *Am. J. Clin. Nutr.*, **27**, 1179
60. Lloyd, J. K. and Muller, D. P. R. (1972). Management of a–β-lipoproteinaemia in childhood. *Protides Biol. Fluids*, **19**, 331
61. Watt, T., Romet, T. T., McFarlane, I., McGuey, D., Allen, C. and Goode, R. C. (1974). Letter. *Lancet*, **ii**, 354
62. Lawrence, J. D., Bower, R. C., Riehl, W. P. and Smith, J. L. (1975). Effects of α-tocopherol acetate on the swimming endurance of trained swimmers. *Am. J. Clin. Nutr.*, **28**, 205
63. Money, D. F. L. (1970). Vitamin E and selenium deficiencies and their possible aetiological role in the sudden death in infants syndrome. *N.Z. Med. J.*, **71**, 32
64. Saltzstein, S. L. (1975). Letter. *Lancet*, **i**, 1095
65. McWhirter, W. R. (1975). Letter. *Lancet*, **i**, 642
66. Khurshid, M., Lee, T. J., Peake, I. R. and Bloom, A. L. (1975). Vitamin E deficiency and platelet functional defect in a jaundiced infant. *Br. Med. J.*, **4**, 19
67. Arimori, S. and Sumitomo, K. (1977). Ultrastructure of platelets of the vitamin E-deficient rats. *J. Nutr. Sci. Vitaminol.*, **23**, 377
68. Steiner, M. (1978). Inhibition of platelet aggregation by alpha-tocopherol. In: *Tocopherol, Oxygen and Biomembranes*. C. de Duve and O. Hayaishi (eds.). pp. 143–163. (Elsevier/North Holland Biomedical Press)
69. Steiner, M. and Anastasi, J. (1976). Vitamin E: An inhibitor of the platelet release reaction. *J. Clin. Invest.*, **57**, 732
70. Lake, A. M., Stuart, M. J. and Oski, F. A. (1977). Vitamin E deficiency and enhanced platelet function: Reversal following E supplementation. *J. Pediatr.*, **90**, 722
71. Machlin, L. J., Filipski, R., Willis, A. L., Kuhn, D. C. and Brin, M. (1975). Influence of

vitamin E on platelet aggregation and thrombocythemia in the rat (38787). *Proc. Soc. Exp. Biol. Med.*, **149**, 275
72. Machlin, L. (1978). Vitamin E and prostaglandins (PG) In: *Tocopherol, Oxygen and Biomembranes*. C. de Duve and O. Hayaishi (eds.). pp. 179–189. (Elsevier/North Holland Biomedical Press)
73. Hope, W. C., Dalton, C., Machlin, L. J., Filipski, R. J., and Vane, F. M. (1975). Influence of dietary vitamin E on prostaglandin biosynthesis in rat blood. *Prostaglandins*, **10**, 557
74. James, P. (1978). Letter. *Br. Med. J.*, **1**, 1345
75. Report. (1976). Vitamin E deficiency and thrombocytosis in Caffey's disease. *Arch. Dis. Child.*, **51**, 393
76. Zannos-Mariolea, L., Tzortzatou, F., Dendaki-Svolaki, K., Katerellos, Ch., Kavallari, M. and Matsaniotis, N. (1974). Serum vitamin E levels with β-thalassaemia major: Preliminary report. *Br. J. Haematol.*, **26**, 193
77. Modell, C. B., Stocks, J. and Dormandy, T. L. (1974). Letter. *Br. Med. J.*, **2**, 259
78. Carpenter, M. P. (1972). Vitamin E and microsomal drug hydroxylations. *Ann. N.Y. Acad. Sci.*, **203**, 81
79. Alberts, D. S., Peng, Y-M. and Moon, T. E. (1978). Alpha-tocopherol pretreatment increases adriamycin bone marrow toxicity. *Biomedicine*, **29**, 189
80. Farrell, P. M. and Bieri, J. G. (1975). Megavitamin E supplementation in man. *Am. J. Clin. Nutr.*, **28**, 1381
81. Corrigan, J. J. Jr. and Marcus, F. I. (1974). Coagulopathy associated with vitamin E ingestion. *J. Am. Med. Ass.*, **230**, 1300
82. Briggs, M. H. (1974). Letter. *Lancet*, **i**, 220
83. Herbert, V. (1977). Letter. *Nutr. Rev.*, **35**, 158

13
The effects of processing on the stability of vitamins in foods

A. E. BENDER

Foods are subjected to a variety of processes which can be broadly classified as dry heat, wet processing (which includes domestic cooking), cooling and chemical additions. On this basis it becomes possible to predict to some extent the stability of vitamins based on their chemical properties such as water solubility, heat lability, light sensitivity etc., but only to a limited extent since many other factors can upset the predictions.

It must be borne in mind that conditions of processing which would provide maximum retention of nutrients are rarely the same as those conferring maximum palatability or keeping properties. Nutrient retention is rarely a primary consideration although modern developments designed to improve the acceptability of processed foods tend to be milder and consequently less damaging to nutrients (by accident if not by design).

STABILITY OF THE VITAMINS

The stability of the vitamins has been investigated in pure solutions in laboratory preparations, in their complex combinations in foods examined under controlled laboratory conditions and under practical conditions with varying degrees of control ranging up to completely uncontrolled conditions such as exist in domestic food preparation.

Consequently the literature contains many apparently contradictory results but some of these, though certainly not all, are due to differences in conditions[1].

Vitamin A

Since retinol and the carotenoids are not soluble in water they do not suffer any losses by extraction into processing and cooking water. In the pure state they are highly susceptible to oxidation but they occur in foods in solution in fats protected by natural antioxidants. Their oxidation depends on the rate of oxidation of the fats themselves since they are attacked by peroxides and free radicals formed from fats. Destruction therefore depends on temperature and access of air, and is promoted by light, traces of iron and especially copper. Both retinol and the carotenes are stable to pH change except that there is some degree of isomerization from all-*trans* to the biologically less potent *cis*-forms at pH 4.5 or less.

Some examples of losses are given in Table 1. De Ritter[2] and Maqsood *et al.*[3] reported that boiling water destroyed 16% of the vitamin A in margarine in 30 min, 40% in 1 h and 70% in 2 h. Frying at 200 °C destroyed 40% in 5 min, 60% in 10 min and 70% in 15 min. Braised liver with an internal temperature of 76 °C lost 0–10% of the retinol[4].

Table 1 Loss of retinol and carotene during storage (data compiled by De Ritter[2])

	Time of storage (months)	Temp. (°C)	Loss (%)
	RETINOL		
Butter	12	5	0–30
	5	28	35
Margarine	6	5	0–10
	6	23	0–20
Skim milk powder	3	37	0–5
	12	23	10–30
Fortified cereal	6	23	20
Fortified potato chips (crisps)	2	23	0
	CAROTENE		
Margarine	6	5	0
	6	23	10
Fortified lard	6	5	0
	6	23	0
Dried egg yolk	3	37	5
	12	23	20
Carbonated beverage	2	30	5
Canned juices	12	23	0–15

Reports of the stability of carotene in canned foods before 1971 are unreliable for it was only then that Sweeney and Marsh[5] showed that carotenoids isomerize during heat processing to isomers of lower biological potency. The relative potency of these isomers is as follows:-

all-*trans*-β-carotene	100
neo-β-carotene-B	53
all-*trans*-α-carotene	53
neo-β-carotene-U	38
neo-α-carotene-B	16
neo-α-carotene-U	13

Most of the carotenoids in fresh vegetables are all-*trans* isomers and are converted during heat processing, such as canning; all-*trans*-β-carotene is partly converted into neo-β-carotene-U with only 38% of its activity. Earlier workers extracted and determined *total* carotenoid pigments and reported no change. Sweeney and Marsh[5] showed that green vegetables containing mainly β-carotene, lost 15–20% of their vitamin A activity, and yellow vegetables, containing mainly α-carotene, lost 30–35% after freezing or canning and subsequent cooking. There appears to be no difference, despite the different temperatures and cooking times, between commercial canning, pressure cooking and conventional cooking.

Losses by air-drying of vegetables and fruits can range from 10–20% under controlled, mild conditions, to almost complete destruction in traditional open-air drying.

The problem of the stability of retinol and carotene added to foods in a concentrated form has been largely solved by the use of stabilized free-flowing powders in which the retinol or carotene is prepared in a starch-coated matrix of gelatin and sucrose, together with added antioxidant.

Thiamin

Second only to vitamin C, thiamin is the most labile of the vitamins, being unstable at neutral and alkaline pH, catalysed by metallic ions such as copper, and completely destroyed by sulphur dioxide, which is often added to foods as a preservative. The greatest losses occur through leaching into the cooking water. Dwivedi and Arnold[6] have reviewed the chemistry of the breakdown and Mulley *et al.*[7] have reviewed the kinetics of the reaction. The latter showed that thiamin is more stable in foods than in model buffer solutions, which they attributed to the protective action of amino acids and proteins and through absorption on starch. For example the addition of cereals to pork has a stabilizing influence on the thiamin[8][9]. The three forms in which it occurs in foods — free thiamin, pyrophosphate and protein-bound — differ in their stability.

The principal loss of thiamin is due to its solubility, the amount lost depending largely on the surface area. Chopped and minced foods can lose 20–70% of their thiamin which is recovered if the extracted liquor is consumed. For example, there is considerable loss from meat into the exuded juices but no destruction at temperatures up to 105 °C; at 200 °C about 20% of the thiamin is oxidized.

The effect of alkalinity is demonstrated by the work of Roy and Rao[10] on losses of thiamin from rice on cooking. There was no loss on boiling in distilled water, 8–10% loss in tap water and 36% loss in well water, so demonstrating that losses were not due to leaching. Alkaline baking powders can cause losses up to 50% on baking, whereas the baking process itself destroys only 15–25%. Sulphur dioxide is sometimes used as a preservative in minced meat (particularly in Great Britain) which leads to very rapid destruction of thiamin — 90% in 48 h[11]. Destruction of thiamin is slow at pH 3, very rapid at pH 5 and immediate at pH 6. Sulphur dioxide may also be used as a preservative in fruit juices although, since there is little thiamin in fruit, the losses are unimportant and, on the contrary, it preserves the vitamin C.

The only vegetable that makes a significant contribution to the intake of thiamin is the potato, which contributes as much as 15% of the daily intake in most countries in Western Europe. Some thiamin is extracted into the processing water depending on the surface-to-volume ratio but modern technology has introduced ready-peeled potatoes and potato chips both to the caterer and domestic user and this necessitates the addition of sulphite to retain the white colour. Oguntona and Bender[12] showed that thiamin is leached out of the deeper as well as the outer layers of the tuber and that losses from a layer 10 mm below the cut surface were 20% in tap water and 55% in sulphite solution (commercially the potato is only dipped in sulphite solution, whereas experimentally it remained immersed in the solution). Furthermore, on subsequent frying, there was an additional loss of 10% of the thiamin from untreated potatoes and 20% from the samples soaked in sulphite. In commercial practice losses can reach 24% after 3 days storage at 5 °C after dipping in sulphite solution; frying losses were 15% untreated and 30% after sulphiting[13].

Baking results in 15–30% loss from bread, mostly in the crust, but thiamin is stable in the loaf after baking. Toasting destroys thiamin to an extent varying with time, thickness of slice and moisture content; in one experiment 10–30% was destroyed in 30–70 s[14]. There is considerable loss of thiamin in dry processing of meat. Canning results in 15–30% loss while the other B vitamins appear to be stable. Dehydration resulted in 25% loss compared with only 8% loss of niacin and no loss of riboflavin: 40% of the thiamin was lost from pre-cooked frozen poultry and 70% loss on canning (Table 2).

Substances that destroy thiamin occur naturally in plants, for example 3, 4-dihydroxycinnamic acid in fern and blueberries, and chlorogenic acid, pyrocatechins and dihydroxycinnamic acid in coffee. Phenol derivatives with *ortho*-hydroxy groups have marked antithiamin activity, those with *meta*-hydroxy groups have medium activity and those with *para*-hydroxy groups are inactive. Polyphenol oxidase catalyses the degradation of thiamin by various plant phenols. There are thiaminases in various fish and crustaceans. Clearly, heat processing that destroys these enzymes has a beneficial effect on the stability of the thiamin.

Riboflavin

Riboflavin is relatively stable to most treatments except exposure to light and, of course, it is leached out of chopped foods in wet processing and cooking. It is stable to oxygen and acid, heat alone does no damage even at 130 °C but it is destroyed under alkaline conditions. Light at acid and neutral pH converts riboflavin to lumichrome, and at alkaline pH to lumiflavin. The reactions are

Table 2   Losses of thiamin in cooked meat (data compiled by Farrer[14])

		% loss	
Beef:	roast	40–60	
	broiled	50	
	stewed	50	(up to 70)
	fried (variable conditions)	0–45	
	braised	40–45	
	canned (85 min at 121 °C)	80	
Pork:	braised	20–30	
	roast	30–40	
Ham:	baked	50	
	fried	50	
	broiled	20	
	canned	50–60	
Chop:	braised	15	
Bacon:	fried	80	
Mutton:	broiled chop	30–40	
	roast leg	40–50	
	stewed lamb	50	
Poultry:	roast chicken, turkey	30–45	
Fish:	fried	40	

irreversible and temperature dependent; they appear to be first order reactions. Milk is the food mostly affected; the lumiflavin formed destroys the vitamin C in milk and even small losses of riboflavin, about 5% can lead to large losses of vitamin C, up to 50%.

This loss in milk has been known for many years but has more recently become of interest with sale of milk in fibreboard cartons under strip lighting in supermarkets. An off-flavour can be detected after 1–15 h in fibreboard cartons compared with 20 min in clear glass bottles. The destruction of riboflavin and vitamin C is directly related to the radiant power emitted between wavelengths 400 and 550 nm but the same relation does not hold for the development of the off-flavour[15]. It is current practice to protect cartons with a layer of aluminium foil. The effect of sunlight on milk in glass bottles gives rise to an unpleasant flavour termed 'sunlight flavour'. This appears to be due to reaction between methionine and riboflavin and to oxidation of the fat. When riboflavin is subjected to light in the presence of oxygen the coagulating time of the milk proteins is increased. It had also been shown that the vitamin C is more stable in coloured bottles — complete destruction of vitamin C took place in 2 h in clear glass bottles, reduced to 10% in dark green glass and 4% in brown glass. Despite this milk seems invariably to be sold in clear glass bottles.

Light has also been shown to affect the riboflavin in small bread rolls. Exposure for 24 h destroyed 17% of the riboflavin, reduced to 13% by wrapping in amber plastic film and to 2% in orange plastic[16].

Nicotinic acid

Nicotinic acid is very stable and the only losses are those caused by leaching. It is stable to heat, air and light at all pH values and is not affected by sulphite. In fact processing confers a benefit from the point of view of nicotinic acid since it is present in an unavailable form in many cereals and it is liberated during heating, particularly under alkaline conditions. The unavailable form is bound to polysaccharides and peptides in a complex called niacytin which is not hydrolysed in the digestive tract. In one series of observations it was shown that 77% of the nicotinic acid in wheat flour was in a bound form, and the amount liberated depended on the time of baking; with alkaline baking powder at pH 9.6 it was completely liberated[17]. The classical example of this alkaline-induced release of nicotinic acid is in the preparation of maize tortillas in Mexico where the maize is soaked in lime water overnight before cooking.

There is some loss of nicotinic acid from meat but this can be completely recovered from the juices. For example, Cover et al.[18] found virtually no loss (less than 10%) of nicotinic acid, pantothenate and riboflavin when beef and pork were roasted at 150 °C while the more labile thiamin suffered 30% loss from beef and 20% from pork (with some recovery from the juices). Roasting at the higher temperature of 205 °C (internal temperature 98 °C) caused a greater loss of thiamin from the meat together with charring of the juices so that the total loss was 50% from both types of meat, while that of nicotinic acid and riboflavin was 30% and that of pantothenate 40%.

Nicotinic acid is also stable in meat during storage — pork stored at −18 °C for 8 weeks lost none, whereas 0–10% pantothenate, 10–20% riboflavin and 10–20% thiamin were lost. In the curing of meat no nicotinic acid is lost in dry curing compared with 40% riboflavin and 15% thiamin. In wet curing 20% of nicotinic acid, 40% of riboflavin and 25% of thiamin is lost.

Folic acid

This presents two problems, firstly it is not clear which of the conjugates is available to man, and secondly most of the assays carried out before 1970 are not acceptable. Accordingly there is little reliable information on processing losses.

Folate is relatively stable at slightly acid pH (below pH 5) but the conjugates differ — monoglutamate is stable but tri- and heptaglutamates are unstable to heat under acid conditions. However, if cooking releases available vitamin from the larger polymers there could even be an increase in biological potency. Malin[19] suggested that folate is unstable only in the free form; he showed losses of 0–10% on steam blanching, 20% on pressure cooking and 25–50% on boiling.

Folate can be destroyed by oxidation. The losses from milk sterilized by ultra-high temperature treatment depend on the amount of oxygen present and can vary between 20 and 100%. It is protected by ascorbic acid (which is added for this purpose to the assay medium) and a similar effect seems to occur in foods. When milk is heated there may be no loss of folate since it is protected by the ascorbic acid. Subsequent reheating however, as in the preparation of infant formulae, can lead to progressive destruction when the ascorbic acid has been destroyed[20]. Folates are also sensitive to sunlight, especially in the presence of riboflavin. One comparison showed 30% loss in one year from tomato juice stored in clear glass bottles compared with only 7% in dark glass. The greater losses take place by extraction into the water.

An example of the cumulative losses in processing was provided by Lin *et al.*[21]. Soaking Garbanzo beans *(Cicer arietinum)* for 12 h leached out 5%; blanching in water at 100 °C caused a loss of 20% in 5 min, 25% in 10 min and 45% in 20 min. Sterilization in the can at 118 °C for 30 min destroyed a further 10%. Since folate is relatively stable at slightly acid pH, there was no further loss on storage in the can nor with a longer sterilizing time. Baking destroyed one-third of the folate naturally present in wheat but only 10% of added folate. It was stable in flour during 1 year's storage at 50 °C and not affected by potassium bromate, azodicarbonamide or benzoyl peroxide used as flour improvers. Losses from foods as diverse as vegetables, fruits and dairy produce average 70% of the free folate and 45% of total folate during overall processing and cooking.

Vitamin B$_6$ (pyridoxine)

Problems arise in interpreting changes in vitamin B$_6$ during processing because of alterations in the chemical form together with differential micro-biological assay. For example, pyridoxal in milk is largely changed to the amine during sterilization which affects the response of the organism. Pyridoxine itself is very stable to heat but the amine and aldehyde are more sensitive. However it is less stable in milk during sterilization or drying possibly because of reaction with –SH groups of the proteins. During storage the amine can complex with –SH compounds leading to the formation, for example, of *bis*-4-pyridoxyl disulphide with only 12–23% of the activity of vitamin B$_6$ in rats. Other reactions can take place between vitamin B$_6$ and amino acids and proteins forming Schiff's bases but these are vitamin-B$_6$ active.

There appears to be no destruction during cooking and the three forms of the vitamin are stable to acid, alkali and oxidation, the main loss being through solubility. For example Raab *et al.*[22] found no loss from Lima beans during the sterilization stage of canning but a 20% loss in blanching in water, reduced to 15% in steam. Derse and Teply[23] found that half the frozen vegetables that they examined had lost 20–40% of their vitamin B$_6$ on cooking, greater losses at this stage than with riboflavin and thiamin. There is considerable destruction of vitamin B$_6$ in milk at high temperatures such as in autoclaving. This destruction was the first evidence of the need for B$_6$ by human beings when it was shown that babies living on milk that had been severely heated suffered convulsions, traced to B$_6$ deficiency. The loss is probably due to reaction with –SH groups of protein since B_6 is stable when heated alone.

It is likely that some of the contradictory reports in the literature of the lability of B_6 are due to different temperatures involved. Table 3 shows that there is negligible loss on pasteurization of milk, 20% loss on shorter time sterilization and 50% loss in the longer time in-bottle sterilization. Others have reported losses of 40–60% on sterilization, presumably under more severe conditions.

Table 3   Typical values for the proportion of heat-labile vitamins in raw milk lost during heat treatment (Porter and Thompson[38])

	Thiamin	Vitamin B_6	Vitamin B_{12}	Folic acid	Vitamin C
Pasteurized	<0.10	<0.10	<0.10	<0.10	<0.25
In-bottle sterilized:					
old method	0.35	0.50	0.90	0.50	0.90
new method	0.20	0.20	0.20	0.30	0.60
UHT sterilized	<0.10	<0.10	<0.10	<0.10	<0.25
Evaporated	0.20	0.40	0.80	0.25	0.60
Sweetened-condensed	0.10	<0.10	0.30	0.25	0.25

Vitamin C

The loss of vitamin C by cooking was described many years before there was a chemical method of assay. In 1920 Chick and Dalyell[24] reported that despite a good supply of vegetables 40 out of 64 children in a hospital in Vienna developed scurvy and that the antiscorbutic activity of cabbage, measured biologically, was reduced by 70% when cooked for 20 min at 100 °C and by 90% when cooked for 60 min at 70–80 °C. Since there was a second heating process (still a common culinary practice) it is possible that by the time the food had been eaten there was virtually complete destruction of vitamin C.

Vitamin C can be lost in the food itself through oxidation by ascorbic acid oxidase (hastened by wilting of leafy vegetables or bruising), by leaching into the cooking or processing water and by oxidation (çatalysed by metals, especially copper). It is readily and reversibly oxidized to dehydroascorbic acid which has a thermal half-life at pH 6 of less than 1 min at 100 °C and 2 min at 70 °C irrespective of the presence of oxygen.

Hence vitamin C is by far the most labile of the vitamins, and indeed of all nutrients, and it is impossible to forecast the vitamin C content of the food as finally eaten (apart from a 'guestimate' that all has been destroyed). To compound this difficulty the initial levels in fruits and vegetables can vary considerably with variety, cultivation conditions and handling after harvesting. It is suggested that fruits and vegetables contain reductase systems which control the level of ascorbic acid and that these are damaged when the product is bruised and the oxidizing enzymes can react with the ascorbic acid.

The oxidase has maximum activity at 40 °C and is almost completely inactivated at 65 °C so rapid heating of the food — such as occurs in blanching — serves to protect the vitamin.

Birch et al.[25] pointed out that two reactions appear to be occurring simultaneously during heating — breakdown of the cell structure allowing contact between enzyme and substrate, and destruction of the enzyme. They found that in peas the point of maximum cell disruption and minimum rate of enzyme inactivation i.e. the fastest rate of oxidation of the vitamin, was 50 °C. Leaching losses are also rapid when the cells are disrupted. Enzymic destruction starts as soon as the crop is harvested. For example kale can lose 1.5% of its vitamin C per hour (one-third in 24 h)[26]. Factors that reduce wilting such as high humidity and cool storage conditions reduce the rate of loss. Losses by extraction into processing water will depend on the surface area-to-volume ratio and the volume of water. Much of this material can be recovered from the water (see Table 4).

Table 4 Vitamin C losses in vegetables during household cooking

Method	Vitamin C (%)		
	Destroyed	Extracted	Retained
Green vegetables			
Boiling (long time, much water)	10–15	45–60	25–45
Boiling (short time, little water)	10–15	15–30	55–75
Steaming	30–40	<10	60–70
Pressure cooking	20–40	<10	60–80
Root vegetables (unsliced)			
Boiling	10–20	15–25	55–75
Steaming	30–50	<10	50–70
Pressure cooking	45–55	<10	45–55

Vitamin C is very stable in canned or bottled foods since air is excluded. There is some destruction in the sterilization process and a small loss by oxidation in the following few weeks due to the oxygen in the air space but when this has been used up the vitamin is stable for long periods. If residual oxygen is used up in the electrochemical process of corrosion less vitamin C is oxidized so in this respect plain cans are superior to lacquered cans or bottles[27]. Kefford et al.[28] reported a slow anaerobic loss of vitamin C from pasteurized fruit juice at a rate about one-tenth that of the oxidation occurring during the first few days. However, since this continued for a long period of time the anaerobic loss finally exceeded the oxidation loss. Anaerobic destruction is accelerated by sucrose, fructose and fructose phosphates leading to the formulation of furfural. This reaction is largely independent of pH but slightly increased at pH 3–4.

Fruits rich in anthocyanins appear to lose vitamin C more rapidly – strawberries can lose 40–60% on processing and 4 months' storage at 37 °C;

raspberries and blackcurrants are even less stable. Vitamin C in apple juice is very unstable — Noel and Robberstad[29] reported 50% loss in 4–8 d and 95% loss in 16 d stored at 5 °C. Sulphur dioxide, used as a fruit juice preservative, serves as an antioxidant for the vitamin C but there is a rapid oxidation when the sealed container is opened. Bender[30] found that orange drink preparations lost 30–50% of the vitamin C within 8 d of opening the sealed bottle and as much as 90% after 3–4 weeks (Table 5), whereas the sealed bottles were stable for many months.

Table 5   Percentage loss of vitamin C from fruit squash exposed to air (Bender[30])

	Days exposure			
	8	15	35	40
Stored full	5		10	
Opened	15	30		90
Stored half full	30	60	70	100

Kinetics of vitamin C destruction

Different mechanisms of vitamin C destruction appear to operate in different foods[31]. In addition to oxidation, losses may be due to non-enzymic browning. In this case activation energy increases with decreasing moisture (as happens with enriched wheat flour and corn–soy–milk mixtures) while in other foods, such as dried orange juice, losses are the same in air and in vacuum. The suggestion is that oxidation may be the main cause of vitamin C destruction at low moisture, and browning at high moisture contents. The loss of ascorbic acid from enriched wheat flour and corn–soy–milk was a first order reaction but the loss from carrot flakes was not affected by temperature.

Table 6   Vitamin C content of green beans (mg/100 g), fresh and after canning under two different conditions (from Marchesini et al.[32])

	Ascorbic acid			Dehydroascorbic acid		
	Fresh	Drained beans	Brine	Fresh	Drained beans	Brine
8 min at 124° C						
Cultivar 1	14.0	0	0.9	7.0	2	0.5
Cultivar 9	11.0	5.0	4.0	5.6	3.4	1.0
Mean of 15 cultivars	17.0±5.0	2.5±1.6	3.7±2.2	6.5±2.0	2.8±1.2	1.9±1.1
% initial value		16%	23%		46%	33%
25 min at 116° C						
Cultivar 1	14.0	3.9	3.8	7.0	3.1	2.8
Cultivar 9	11.0	3.4	2.1	5.6	2.1	2.1
Mean of 15 cultivars	17.0±5.0	3.4±1.6	4.1±2.3	6.5±2.0	2.6±1.3	1.8±1.0
% initial value		21%	25%		44%	29%

The difficulty in producing a mathematical model in the present state of knowledge is illustrated by the observations of Marchesini et al.[32] (see Table 6). Even among a range of 15 cultivars of green beans grown under identical conditions there was a considerable difference in the amounts of ascorbic and dehydroascorbic acids, their rates of loss and the amounts leached into processing water. Under the same conditions, namely, canning, 8 min 124 °C, one cultivar lost all its ascorbic acid and three-quarters of the dehydro-ascorbic acid, while another lost only one-quarter of the ascorbic and one-quarter of the dehydro form.

Vitamin D

This vitamin is generally regarded as being very stable but little information is available because of the relatively few assays that have been carried out. It will withstand smoking of fish, pasteurization and sterilization of milk, and spray-drying of eggs, although it is generally considered to be similar to vitamin A in being destroyed in oxidizing fats. When used to enrich infant milks it is common practice, based on a limited number of assays, to allow for the destruction of 25–35% of the added vitamin during the drying process.

Vitamin E

Tocopherols occur in many fats as natural antioxidants and are therefore preferentially destroyed under oxidizing conditions such as exposure to air and light, accelerated by heat and metals such as copper. However such changes are relatively slow and processing losses are small. Little is lost, for example, when vegetable oils are fried. However vitamin E is unique in being unstable during frozen storage. Bunnell et al.[33] showed that while the oil used for frying lost 10% of its vitamin E content in the process, the oil absorbed into the food lost vitamin E rapidly when stored at −12 °C. Potato chips (crisps) lost 48% in two weeks at room temperature, 70% in 4 weeks and 77% in 8 weeks. When stored at −12 °C losses were almost the same. The loss is believed to be caused by the formation from unsaturated fatty acids of hydro-peroxides which are relatively stable at low temperatures whereas they normally decompose, first to peroxides then to aldehydes and ketones, which are less damaging to the vitamin E.

Esters are more stable than free tocopherols. Bunnell et al.[33] showed that the acetate was only 10–20% destroyed under conditions where free tocopherol was completely destroyed. Boiling destroys 30% of the toco-pherol in sprouts, cabbages and carrots and the losses on canning are considerable. However vegetables are not an important source of vitamin E in the diet. Fifty per cent of the vitamin E is destroyed during bread making through the use of chlorine dioxide as a bleaching agent.

PROCESSES

Blanching

Processes such as freezing and drying are preceded by blanching to inactivate enzymes which would cause deterioration during storage even at low temperatures. Canning may also be preceded by blanching to wilt bulky vegetables and reduce their volume, to expel gases, maintain colour and to clean them. Such wet processing causes a loss of water-soluble nutrients although there are modifications designed to minimize such losses. In general, there is little overall difference between vitamin losses incurred in canning and drying since the greatest losses take place at the blanching stage. Losses into processing and cooking water depend on the state of subdivision of the food (surface-area-to-volume ratio) relative volumes of water to food, time and sometimes temperature. The commonest procedure is immersion in hot water but a variety of other methods including hot air, steam and irradiation have been developed. The time of treatment varies with the food, its size and the particular process. For example, the time may vary for brussel sprouts from 3.5 to 7 min depending on size, garden peas can be treated in 1 min; sliced beans may require 2 min and temperatures range from 93 to 99°C[34]. Oxidation can occur as well as leaching out.

Losses are affected also by enzymic destruction. Birch et al.[25] found that inactivation of the enzyme occurred at 85°C and the extracted vitamin C could be recovered completely from the blanching water. At 70°C enzymic action occurred as well as leaching and 24% of the total vitamin C was oxidized in 2 min.

The extraction into treatment water that occurs in blanching, canning and cooking may not be a loss if the aqueous material is consumed. In a blanching process the water is discarded, but canned liquor may be consumed in products such as fruits and meat with gravy, while the salt water from canned vegetables is usually (but not necessarily) discarded. Vitamins extracted into the cooking water of meat are not lost since the gravy is almost always consumed.

Microwave blanching is reported to cause less damage than steam and a combination of microwave and hot water treatment has been claimed to produce a product superior both in nutritional quality and palatability[35]. Another method is fluidized-bed blanching which is effectively hot gas treatment. Losses of vitamin C and also of carotene were reduced in this process.

Frozen foods

The freezing process itself has no effect on the vitamins, apart possibly from a small loss, less than 10%, of the vitamin C, and they are stable in the frozen foods for periods of a year or more. However, freezing must be preceded by blanching with the losses described earlier. Frozen foods present an example of nutrient losses in processing being in place of, rather than in addition to, final cooking, since the processing reduces the final cooking time. As a result there is often little difference between the vitamin content of the final cooked food, whether originally frozen or fresh.

However, fruits and vegetables required for freezing are harvested and processed at the peak of condition which usually means at the maximum vitamin content, while so-called 'fresh foods' are invariably past their peak when purchased. Consequently foods may have a higher vitamin content after freezing and cooking than if cooked fresh. Freezing has no effect on B vitamins but the rate of freezing can influence the subsequent loss of exudate from meat during thawing and cooking. From the limited amount of information available thiamin, riboflavin, niacin and pantothenate appear to be quite stable during frozen storage while there are losses of vitamin B_6[36].

Pressure cooking

There are many conflicting reports in the literature comparing pressure cooking with conventional boiling. Theoretically both the shorter time required and the smaller losses expected in leaching should lead to marked superiority of pressure cooking, and while some authors have shown this to be so, others have found no difference.

The principal cause of loss is by leaching rather than heat and Krehl and Winters[37] found that for four vitamins and two minerals losses were the same when the same volume of water was used despite the marked differences in cooking times. In a number of vegetables it was shown that losses of thiamin in pressure cooking were 25–50%, compared with 50% on steaming and 75–80% on boiling; for vitamin C the relative losses by the different methods varied with the vegetable. Consequently it is not possible to draw conclusions about the relative merits of pressure cooking in this respect.

Microwave heating

Microwave heating with high energy electromagnetic radiation (usually of frequency of 2450 mHz, wavelength 12 cm) is an extremely efficient process. In contrast to conventional cooking, where heat is applied to the surface and then conducted to the inner parts, microwaves generate heat throughout the bulk of the food. The procedure is used in catering where it has the advantage of allowing quick preparation of small quantities of food as required, instead of having to keep a large amount hot with detriment to both palatability and nutrient content. It is used commercially for processes such as 'finish drying' of potato chips and is being widely adopted for domestic use.

Despite the lower energy input, lower surface temperature and shorter cooking time compared with conventional cooking methods, reports of nutrient losses are contradictory, some authors finding it superior while others find no difference. The reason for the apparent contradiction appears to lie in variation with the type of food under examination.

CONCLUSIONS

With very few exceptions it is not possible to forecast the stability of the vitamins to the various treatments involved in food processing. Only a limited number of direct comparisons have been carried out between the starting material and the final processed and cooked food and most of these have involved laboratory preparations treated under controlled conditions unlike those of commercial and domestic preparation.

Fresh foods, both plant and animal, vary considerably in vitamin content, depending on variety, treatment during growth and treatment after cropping. Then the conditions of processing vary greatly so it is not surprising to find that processed foods vary considerably in vitamin content.

Information about the effects of processing is derived from studies of the isolated vitamins in pure form and in the complex mixtures that constitute foods. The effects of light, heat, oxygen and other factors may be examined under controlled laboratory conditions and also under the very varied conditions that operate in practice. Processed foods can be analysed after subjection to the differing conditions operating in different factories and also at the point of sale after the additional hazards of transport and storage under various conditions. These investigations provide the information on which conclusions are based about the stability of the vitamins in foods but they do not take account of the vicissitudes of domestic storage, preparation and other treatments. The stability of the vitamins under these extremely varied and uncontrolled conditions determines the intake and consequently the nutritional status of the consumer but there is very little information indeed available about nutrients 'on the plate'.

References

1. Bender, A. E. (1978). *Food Processing and Nutrition.* (London: Academic Press)
2. De Ritter, E. (1976). Stability characteristics of vitamins in processed foods. *Food Technol.,* **30**, 48
3. Maqsood, A. S., Haque, S. A. and Khan, A. H. (1963). Stability of vitamin A in ghee and vitaminised vanaspati. *Pakist. J. Sci. Ind. Res.,* **6**, 119
4. Kizlaitis, L., Deibel, C. and Siedler, A. J. (1964). Nutrient content of variety meats. II. Effects of cooking on vitamin A, ascorbic acid, iron and proximate composition. *Food Technol.,* **18**, 103
5. Sweeney, J. P. and Marsh, A. C. (1971). Effect of processing on provitamin A in vegetables. *J. Am. Dietet. Assoc.,* **59**, 238
6. Dwivedi, B. K. and Arnold, R. G. (1973). Chemistry of thiamin degradation in food products and model systems. A review. *J. Agric. Food Chem.,* **21**, 54
7. Mulley, E. A., Stumbo, C. R. and Hunting, W. M. (1975). Kinetics of thiamin degradation by heat. *J. Food Sci.,* **40**, 985, 989, 993
8. Rice, E. E., Beuk, J. F. and Robinson, H. E. (1943). The stability of thiamin in dehydrated pork. *Science,* **98**, 449
9. Rice, E. E., Squires, E. M. and Fried, J. F. (1948). Effect of storage and microbial action on vitamin content of pork. *Food Res.,* **13**, 195
10. Roy, J. K. and Rao, R. K. (1963). Alkalinity of cooking water and stability of rice. *Ind. J.*

Med. Res., **51**, 533

11. Hermus, R. J. J. (1970). Sulphite-induced thiamine cleavage. Effect of storage and preparation of minced meat. *Nutr. Abstr. Rev.*, **40**, 51

12. Oguntona, T. E. and Bender, A. E. (1976). Loss of thiamin from potatoes. *J. Food Technol.*, **11**, 347

13. Mapson, L. W. and Wager, H. G. (1961). Preservation of peeled potatoes. I. Use of šulphite and its effect on thiamine content. *J. Sci. Food Agric.*, **12**, 43

14. Farrer, K. T. H. (1955). The thermal destruction of vitamin B_1 in foods. *Adv. Food Res.*, **6**, 257

15. Hedrick, T. I. and.Glass, L. (1975). Chemical changes in milk during exposure to fluorescent light. *J. Milk Food Technol.*, **38**, 129

16. Stephens, L. C. and Chastain, M. F. (1959). Light destruction of riboflavin in partially baked rolls. *Food Technol.*, **13**, 527

17. Clegg, K. M. (1963). Bound nicotinic acid in dietary wheaten products. *Br. J. Nutr.*, **17**, 325

18. Cover, S., Dilsaver, E. M., Hays, R. M. and Smith, W. H. (1949). Retention of B vitamins after large-scale cooking of meat. II. Roasting by two methods. *J. Am. Dietet. Assoc.*, **25**, 949

19. Malin, J. D. (1975). Folic acid. *World Rev. Nut. Diet.*, **21**, 198

20. Ford, J. E., Porter, J. W. G., Scott, K. J. *et al.* (1974). Comparison of dried milk preparations for babies on sale in 7 European countries. (ii). Folic acid, vitamin B_6, thiamin, riboflavin and vitamin E. *Archs. Dis. Child.*, **49**, 874

21. Lin, K. C., Luh, B. S. and Schweigert, B. S. (1975). Folic acid content of canned Garbanzo beans. *J. Food Sci.*, **40**, 562

22. Raab, C. A., Luh, B. S. and Schweigert, B. S. (1973). Effect of heat processing on the retention of vitamin B_6 in Lima beans. *J. Food Sci.*, **38**, 544

23. Derse, P. H. and Teply, L. J. (1958). Effect of storage conditions on nutrients in frozen green beans, peas, orange juice and strawberries. *J. Agric. Fd. Chem.*, **6**, 309

24. Chick, H. and Dalyell, E. J. (1920). The influence of over-cooking vegetables in causing scurvy among children. *Br. Med. J.*, Oct. 9

25. Birch, G. G., Bointon, B. M., Rolfe, E. J. and Selman, J. D. (1974). Quality changes involving vitamin C in fruit and vegetable processing. In: *Vitamin C.* G. G. Birch and K. J. Parker. (eds.). (London: Applied Science Publishers Ltd.)

26. Zepplin, M. and Elvehjem, C. A. (1944). Effect of refrigeration on retention of ascorbic acid in vegetables. *Food Res.*, **9**, 100

27. Adam, W. B. (1941). *Rep. Fruit Veg. Preserv. Res. Stn.*, **14**

28. Kefford, J. F., McKenzie, H. A. and Thompson, P. C. (1959). Effect of oxygen on quality and ascorbic acid retention in canned and frozen orange juices. *J. Sci. Food Agric.*, **10**, 51

29. Noel, G. L. and Robberstad, M. T. (1963). Stability of vitamin C in canned pineapple juice and orange juice under refrigerated conditions. *Food Technol.*, **17**, 947

30. Bender, A. E. (1958). The stability of vitamin C in a commercial fruit squash. *J. Sci. Food Agric.*, **11**, 754

31. Labuza, I. (1972). Nutrient losses during drying and storage of dehydrated foods. *CRC Critical Reviews of Food Technology*, **3**, 217

32. Marchesini, A., Majorino, G., Montuori, F. and Cagna, D. (1975). Changes in the ascorbic and dehydroascorbic acid content of fresh and canned beans. *J. Food Sci.*, **40**, 665

33. Bunnell, R. H.; Keating, J., Quaresimo, A. and Parman, G. K. (1965). Alpha-tocopherol contents of foods. *Am. J. Clin. Nutr.*, **17**, 1

34. Bomben, J. L., Dietrich, W. C., Hudson, J. S. *et al.* (1975). Yields and solids loss in steam blanching, cooling and freezing vegetables. *J. Food Sci.*, **40**, 660

35. Dietrich, W. C., Huxsell, C. C. and Guadagni, D. G. (1970). Comparison of microwave, conventional and combination blanching of brussels sprouts for frozen storage. *Food Technol.*, **24**, 613

36. Kotschevar, L. H. (1955). B-vitamin retention in frozen meat. *J. Am. Dietet. Assoc.*, **31**, 589

37. Krehl, W. A. and Winters, R. W. (1950). Effect of cooking methods on retention of vitamins and minerals in vegetables. *J. Am. Dietet. Assoc.*, **26**, 966

38. Porter, J. W. G. and Thompson, S. Y. (1976). Effects of processing on the nutritive value of milk. *Proc. 4th Int. Cong. Food Sci. Technol. Madrid*, **Vol. 1**

14

The clinical diagnosis of vitamin deficiencies in everyday medical practice

A. N. EXTON-SMITH

Primary dietary deficiency of vitamins is now rare in Britain except in certain deprived sections of the population. In infancy and childhood vitamin D deficiency leading to rickets is seen mainly in coloured immigrants. At the other end of life nutritional deficiency of vitamins occurs as a result of the changes in bodily systems associated with senescence and the altered socio-economic circumstances of old people. Much more common, however, are the secondary deficiencies due to disease processes impairing appetite and the absorption, metabolism and utilization of vitamins. Although the diseases responsible for these impairments can occur at all ages they are much more frequent in the elderly who, in addition, are more likely to suffer from disorders affecting several organs of the body.

Malnutrition can be defined as a disturbance of form or function due to lack of energy or of one or more nutrients[1]. A diagnosis cannot be made on the sole basis of a low dietary intake or of the presence of abnormal biochemical findings, although these should alert the clinician to the possibility of nutritional deficiencies. In the Nutrition Survey of the Elderly[1] the results based on random samples of old people in six areas of the UK showed that malnutrition of all forms occurs in about 3% of the elderly population. But the clinical significance of malnutrition is far greater than this prevalence might suggest since in almost every case it is treatable with excellent results. Moreover, there may be a much bigger problem of subclinical malnutrition owing to the long latent period before clinical manifestations develop in vitamin deficiency states. In some cases the symptoms and signs may be different from the classical descriptions of the vitamin deficiency diseases; for example, Vitamin D deficiency is associated with an increased liability to fractures in the absence of the typical features of rickets or osteomalacia. The most common types of vitamin malnutrition encountered in clinical practice are those associated with deficiencies of the B group, ascorbic acid and vitamin D. In the following description attention will be directed mainly to their occurrence in two vulnerable sections of the population, namely, childhood and old age.

VITAMIN B COMPLEX DEFICIENCY

Clinical features

Deficiency of the B group vitamins leads to changes in the mucous membranes, the heart and the nervous system. In the mouth and tongue the signs include:

Cheilosis
Red, denuded, often scaly, epithelium at the line of closure of the lips;

Angular stomatitis
Greyish white, sodden and swollen epithelium, progressing to fissuring, radiating outwards from the corners of the mouth;

Naso-labial seborrhoea
Enlarged follicles around the sides of the nose, and plugged with sebaceous material;

Glossitis
Bare, red, smooth tongue with loss of filiform papillae, sometimes associated with fissuring and enlargement of the fungiform papillae.

A high incidence of these changes, which have been attributed to deficiency of riboflavin, nicotinic acid and possibly pyridoxine, have been reported in elderly patients by some authors[2-4].

Brin[5][6] has proposed five stages in the development of vitamin deficiency disease based on observations made on the development of human thiamin deficiency. It is noteworthy that the overt disease state is not manifest until stage four. During the 'physiological' stage, three of the symptoms are loss of appetite, general malaise and increased irritability—all common complaints due to many other causes, especially in the elderly. In stages four and five thiamin deficiency leads to cardiac and neurological manifestations including bradycardia, cardiac enlargement, oedema, peripheral neuropathy, mental confusion and ophthalmoplegia. Wernicke's encephalopathy is almost certainly due to thiamin deficiency; the clinical features are due to changes in the central and autonomic nervous systems and include diplopia, nystagmus, postural hypotension, ophthalmoplegia and the mental disorders of Korsakoff's psychosis — loss of memory, disorientation, confabulation and hallucinations. In Britain and the United States it is usually associated with alcoholism, but Philip and Smith[7] have described Wernicke's encephalopathy in old people with accidental hypothermia and it is possible that the poor nutritional status of many people who are admitted with hypothermia is responsible. Acute confusional states associated with toxic-infective processes, which account for one-third of cases of mental disorder in old people admitted to hospital[8] may in some instances also be attributable to a relative thiamin deficiency. In these patients blood pyruvate levels are often elevated and fall again when the underlying process (for example, pneumonia) responds to appropriate treatment. These observations need to be confirmed by serial red cell transketolase estimations and it is possible that the development of the acute confusional state is related to the initial status of thiamin nutrition.

Diagnosis

In the patient who has cheilosis, angular stomatitis, naso-labial seborrhoea and glossitis, riboflavin deficiencies should be suspected. There are, however, more common causes of some of these conditions; angular stomatitis is more often due to ill-fitting dentures, glossitis can be due to deficiency of iron, nicotinic acid, folic acid and vitamin B_{12}. Confusional states are due to very many other causes besides thiamin deficiency. Nevertheless it is important that a firm diagnosis should be made since all these conditions respond well to treatment with appropriate vitamin supplementation. Wernicke's encephalopathy in particular responds dramatically to the administration of thiamin, but if left untreated this condition is usually fatal.

Thus in all cases of suspected vitamin B complex deficiency the clinical diagnosis must be confirmed by biochemical tests. Brin and his colleagues[5][6] have assessed thiamin status by determining its urinary excretion level by measurement of the erythrocyte transketolase (TK) activity and by calculation of the TPP effect. In the last test thiamin pyrophosphate (TPP), the coenzyme of thiamin, is added to the haemolyzed cells and the TPP effect is calculated from the increment of TK activity due to the addition of TPP and expressing this as a percentage of the TK activity of unenriched blood. On a thiamin depleted diet urinary thiamin excretion is reduced to 50 μg daily after about 5 d; the TK activity is depressed with a positive TPP effect of about 15% and the urinary thiamin is reduced to 25 μg after about 10 d, the third or 'physiological' stage is reached by 21 d when the urinary thiamin is 0.25 μg daily and the TK activity is reduced 15–25% with a TPP effect up to 30%. By the time clinical manifestations of deficiency appeared the TK activity was reduced more than 35% and the TPP effect was in excess of 40%. It was noted that the urinary thiamin excretion reached a minimum level in 12 d and once this level was reached it was not possible to differentiate between a preclinical deficiency state and severe disease by this measure alone. Values of TPP effect below 15% were taken as normal; above this critical level there was found to be a definitive curve of relationship between TK activity and TPP effect. In a study of 233 subjects aged 44–94 years Brin et al.[9] found thiamin deficiency in 37% as judged by reduced thiamin excretion and in 5% if based on a TPP effect of greater than 15%.

The riboflavin status of 128 old people has been investigated by Thurnham[10] in a study of accidental hypothermia based on a random sample of the elderly population in the London Borough of Camden. The erythrocyte glutathione reductase activity (EGR) and the percentage stimulation of EGR by flavin adenine dinucleotide (FAD) were measured. A percent stimulation of greater than 30% (usually regarded as the upper limit of normal) was found in 18% of the men and in 19% of the women. Thus it is considered that there may be marginal riboflavin deficiency in about one-fifth of the elderly population. Although the true significance of this marginal deficiency is at present unknown these laboratory tests serve to identify those individuals who might benefit from vitamin supplementation to prevent the appearance of acute deficiency disease.

FOLATE DEFICIENCY

Clinical features

Although folate deficiency causes a general disturbance in which various tissues and organs are involved the main clinical findings are anaemia and changes in the nervous system and mucous membrane of the tongue. The anaemia is megaloblastic characterized by abnormal nucleated red cell precursors (megaloblasts) in the bone marrow and macrocytes in the peripheral blood. There is associated leukopenia with hypersegmentation of the neutrophils and the number of platelets may be reduced. There may be the complaint of soreness or burning of the tongue and the surface may be red due to acute glossitis; but it is more commonly smooth, shiny and atrophic. In patients with chronic neurological diseases, especially peripheral neuropathy, the changes including megaloblastic anaemia, are more often due to folate deficiency than to vitamin B_{12} deficiency[11]. Mental changes may precede the anaemia; they are non-specific and include mild confusion, depression, apathy and intellectual impairment. Cases of dementia due to folate deficiency have been reported and Strachan and Henderson[12] noted a response in two patients with dementia to prolonged folate administration.

Diagnosis

Herbert[13] maintains that folate deficiency is the commonest vitamin deficiency in man. Although a folate-free diet quickly leads to lowering of serum folate levels, many months must elapse before clinical and haematological changes develop. Folate deficiency should be suspected in the presence of macrocytic anaemia, glossitis and unexplained peripheral neuropathy. A search should be made for possible causes. In addition to reduced dietary intake, folate deficiency may be due to malabsorption, increased utilization and impaired effectiveness. Absorption of folate occurs in the upper small intestine and it may be impaired in gluten enteropathy and in other conditions giving rise to malabsorption syndrome. Folate requirements are increased in pregnancy and when cell turnover increases e.g. in haemolytic anaemia, myeloproliferative syndromes, carcinoma, myeloma and in certain chronic inflammatory diseases such as tuberculosis and Crohn's disease. Under certain conditions the effectiveness of available folate is impaired e.g. vitamin C deficiency inhibits folate coenzymes and in scurvy impaired folate metabolism can lead to megaloblastic anaemia (*vide infra*); various drugs including methotrexate, anticonvulsants and trimethaprim interfere with folate metabolism. Folate status should be assessed by measurement of serum and red cell folate levels: deficiency should be suspected when the serum level is less than 3 ng/ml and the red cell concentration less than 150 ng/ml. In the Nutritional Survey of the Elderly[1] serum folate levels of less than 3 ng/ml were found in 14.6% of the subjects living at home. Red cell folate levels of less than 150 ng/ml were found in 16.1% indicating chronic folate deficiency; more severe degrees of deficiency with red cell folate concentrations of less than 100 ng/ml were found in 3.7% of the subjects studied. There was a significant correlation between serum and red cell folate concentrations ($p<0.01$). In elderly patients admitted to hospital there is a higher incidence of folate deficiency. Batata *et al.*[14] in Oxford found that 10% of patients over the age of 60 had serum folate levels of less than 2.1 ng/ml. A nutritional origin was suspected since with severe disability (and in consequence an inability of the patient to look after himself) the more likely was there to be folate deficiency; there was found to be a statistically significant relationship between organic brain disease and low folate levels.

ASCORBIC ACID DEFICIENCY

Clinical features

The classical manifestations include swelling and bleeding from the gums (not seen in edentulous individuals), weakness, anaemia, extensive haemorrhages in the skin of the legs and arms ('sheet' haemorrhages) and sometimes haemorrhages at other sites. Russell *et al.*[15] have shown that vitamin C deficiency may play a part in maintaining gastrointestinal haemorrhage which had been precipitated initially by gastric irritants such as aspirin. Sublingual 'petechiae' have been regarded by Taylor[3] as an early sign of scurvy, but Andrews *et al.*[16] have shown by histological examination that the lesions are small aneurysmal dilatations of the minute vessels under the tongue. It is highly unlikely, therefore, that they can be caused by acute vitamin C deficiency; moreover, they do not disappear when ascorbic acid intake is increased.

Anaemia is common in scurvy; it is usually normocytic or macrocytic with normoblastic or macronormoblastic erythropoiesis, but cases of true megaloblastic anaemia have been reported[17][18]. The origin of anaemia is often multi-factorial: haemolysis, bleeding, dietary deficiency of iron and derangement of red cell metabolism have been incriminated[19]. Megaloblastic change has been attributed either to an associated dietary folate deficiency or to impairment of folate metabolism in scurvy. Stokes *et al.*[20] have shown that megaloblastic anaemia of scurvy is caused in part by removal of tetrahydrofolates from the metabolic pool due to oxidation to 10-formyl folic acid. It is believed that an important role of ascorbic acid is to prevent the oxidation of the tetrahydrofolates and thus to maintain the availability of the folate metabolic pool.

Mental changes are probably common in scurvy. Walker[21] drew attention to depression associated with vitamin C deficiency and Kinsman and Hood[22] to personality changes in experimentally induced scurvy.

In human volunteers fed on a diet completely lacking in vitamin C these changes appeared after about 30 d, long before there were any classical manifestations of scurvy. The personality changes produced decrements in psychomotor performance associated with reduced arousal and motivation. It has been known for many years that vitamin C may be concerned with wound healing. Most of the evidence is derived from animal experiments; thus it has been shown that an increased vitamin C intake in guinea pigs produces an increased wound strength in healing skin. The evidence from human experiments is less direct. Burr and Rajan[23] administered 1 g ascorbic acid daily for 3 d to seven paraplegic patients suffering from pressure sores. Biopsy of the skin around the pressure sores was performed before and after treatment. More intense staining for collagen was seen in the second biopsy specimen, but no changes were observed following the administration of a placebo. Thus vitamin C promotes collagen formation, but this is not necessarily equated with better wound healing.

Recently Irvin et al.[24] have investigated the ascorbic acid requirements in postoperative patients. In 63 patients who had been submitted to surgical operation involving minor to major trauma, there was a significant reduction in leukocyte ascorbic acid levels following operation. The postoperative changes were unrelated to the extent of surgical trauma but there was a significant correlation between the postoperative ascorbic acid measurements and white cell count. The authors suggest that the postoperative leukocytosis and release by the bone marrow of leukocytes with a low ascorbic acid content may partly account for the postoperative changes in leukocyte ascorbic acid measurements. Nevertheless, as the authors point out, surgical operations were followed by an authentic increase in ascorbic acid requirements as shown by a 42% reduction in the leukocyte ascorbic acid levels on the third postoperative day. The results of these studies support the argument for the use of ascorbic acid supplements in surgical patients, although the benefits of supplementation have not been assessed.

Diagnosis

When the typical haemorrhagic manifestations of scurvy are present the diagnosis can often be suspected especially when the patient is an elderly man who has recently been widowed and who must fend for himself. If he lives mainly on convenience foods, tea, bread and butter and jam, the diet will contain very little vitamin C. The results of several nutrition surveys, for example, Milne et al.[25], DHSS[1] disclose that there is a significant number of old people whose intake of vitamin C is less than 10 mg/d, which is known to be the amount required to prevent or cure scurvy[26]. The recommended allowance of 30 mg daily[27] takes into account the considerable individual variations in requirements and increased requirements due to stress. Although a high proportion of the elderly population are consuming less than 30 mg/d the majority will not suffer any ill-effects.

In suspected cases of scurvy the vitamin C status should be assessed by measurement of the leukocyte ascorbic acid (LAA). Our assessment of vitamin C requirements is handicapped by lack of knowledge of the tissue levels needed for health. Windsor and Williams[28], by measuring total hydroxyproline excretion (THP) in response to vitamin C, found that THP increased when the initial LAA content was less than 15 μg/10^8 leukocytes, but when the LAA level was higher than this the response to vitamin C supplementation failed to occur. Thurnham[10] in an investigation of old people who participated in the Camden Survey of deep body temperatures in the elderly[29] found that 28% of the men and 10% of the women had LAA levels of less than 15 μg/10^8 leukocytes. The higher proportion of men with deficiency is in keeping with the findings of clinicians that scurvy is more prone to occur in men than in women.

VITAMIN D DEFICIENCY

Deficiency of vitamin D leads to a generalized disorder of bone — rickets in childhood and osteomalacia in adult life. Each condition is characterized by a deficient calcification of a normal bone matrix. Histological examination reveals an excessive amount of osteoid around the bone trabeculae. In addition, recent work has shown that vitamin D deficiency is of importance in the pathogenesis of certain fractures in the absence of the typical features of rickets and osteomalacia.

Clinical features

Rickets

Active rickets occurs most frequently at times of rapid growth, namely, in the rapidly growing infant and toddler and during the pubertal growth spurt. It may occur shortly after birth if the mother is vitamin D deficient. The child appears restless and fretful, the muscles are weak and hypotonic, and abdomen is protruberant due to the downward displacement of the diaphragm by the deformed chest. The forehead is enlarged and square with a 'hot-cross bun' due to bossing in the parietal and frontal regions and failure of the anterior fontanelle to close. Swellings due to excess osteoid tissue at the junction of the bone and cartilage in the ribs give rise to 'rachitic rosary' or 'beading'. The softened bones bend giving rise to deformity of the chest, the limbs and the pelvis. There is swelling of the wrist and ankles and enlargement of the epiphyses. Sometimes increased irritability of the nervous system, due to lowered serum calcium, is a striking feature, and the manifestations include carpo-pedal spasm, laryngismus stridulus and convulsions.

Osteomalacia

In the early stages the patient often complains of vague pains in a variety of sites and these are usually dismissed as 'rheumatism'. Later the pain becomes more persistent and localized to the bones which are extremely tender to pressure. The patient becomes shorter owing to deformation of the trunk — usually kyphosis — and on radiography the intervertebral discs are often ballooned and the soft vertebral bodies become biconcave ('cod-fish vertebrae'). Other bones may become deformed such as the sternum, the pelvis and the femoral necks. Another characteristic finding on radiographic examination is the appearance of Looser's zones or pseudofractures. These are bands of decalcification perpendicular or oblique to the surface of the bone and on either side of the translucency there may be a denser band of callus. They occur in only about one-third of cases of osteomalacia, but as they are pathognomonic of osteomalacia, a skeletal survey should be conducted in suspected cases, particular attention being paid to the pubic rami, the femoral neck, the ribs, the border of the scapula, the upper end of the humerus and less frequently the shafts of the tibia, fibula, radius and ulna. In addition to Looser's zones true spontaneous fractures occur.

Muscular weakness is often striking, and in osteomalacia it usually takes the form of a proximal myopathy affecting the muscles of the pelvic and shoulder girdles. The patient may complain of difficulty in climbing stairs and in getting up from a chair; when walking the patient may have a typical 'waddling gait'. When the shoulder girdle muscles are involved the patient is unable to raise the arms and perform such activities as brushing the hair.

Fractures in the absence of the typical features of rickets and osteomalacia

After the First World War, from 1918 to 1922, when margarine, which at that time contained no vitamin D, replaced butter a higher incidence of fractures in children was observed. In a prosperous and well-endowed public school the incidence of fractures occurring from ordinary activities of school life and games varied from seven a year when butter was used to eighteen a year on its replacement by margarine. In older adults Aaron et al.[30] in Leeds have shown by histological methods that 20–30% of women with fracture of the proximal femur and about 40% of men have bone changes characteristic of osteomalacia. Later they showed[31] that the proportion with osteomalacia varied with the season. The highest frequency of abnormal calcification fronts (43%) was observed in February–April and lowest (15%) in August–October. The highest frequency of abnormal osteoid covered surfaces (47%) was observed in April–June and the lowest (13%) in October–December. They concluded that variation in hours of sunshine is responsible for a seasonal variation in osteomalacia in femoral neck fractures and, possibly, in the elderly population as a whole. The significance of vitamin D deficiency as an important factor in the pathogenesis of fracture of the femoral neck has been confirmed by the study of Faccini, Exton-Smith and Boyde[32]. The mean value of trabecular osteoid area in the fracture group was 4% compared with 1% in a control group. The difference was also striking in the proportion of trabecular surface covered by osteoid; the mean value for the fracture group was 24.5% compared with 7.9% for the control group. Brown *et al.*[33] found significantly lower levels of 25-hydroxycholecalciferol in patients with fractures of the femoral neck compared with those in controls of similar age from whom blood samples were taken at the same time of year. This is believed to be a reflection of the decreased out-of-doors activity of the patients prior to their fracture.

Diagnosis

In the active phase of rickets the clinical diagnosis can usually be confirmed by finding the typical changes in bone on skeletal X-ray, together with a lowered serum calcium and inorganic phosphorus and a raised alkaline phosphatase. In the adult, osteomalacia can be diagnosed with certainty if Looser's zones are present at characteristic sites in the skeletal radiographs. But in many cases Looser's zones are absent and the condition has to be differentiated from osteoporosis. In each disease skeletal rarefaction and crush fractures of the vertebral bodies occur but in osteoporosis the bones are brittle and do not bend, whereas in osteomalacia they are soft and tender on pressure. When osteoporosis is complicated by fracture a rise in the serum alkaline phosphatase may occur and the biochemical changes may be confused with osteomalacia. The finding of a greatly reduced serum 25-hydroxycholecalciferol (less than 8 ng/ml) may be of help in the diagnosis of osteomalacia. There are, however, many suspected cases in which the diagnosis remains in doubt and bone biopsy will be necessary so that histological examination can be carried out on undecalcified sections. Diagnosis may also be confirmed by a therapeutic trial with vitamin D. Monitoring of treatment should be undertaken by such procedures as demonstrating the rise in total hydroxyproline excretion in response to vitamin D[34], the decreased proportion of osteoid with a calcification front and its increase with therapy[35] and the rise in serum inorganic phosphorus in response to intravenous vitamin D_3[36].

When the diagnosis of rickets or osteomalacia is confirmed a search must be made for the cause. It may be due to primary vitamin D deficiency associated with dietary lack or inadequate synthesis in the skin resulting from lack of exposure to sunlight. There is evidence that skin synthesis is a more important source than dietary intake (see Lawson, pp 91–100). In this country primary deficiency is most often seen in young Asian immigrants and in housebound old people who have dietary inadequacy as well as decreased ultraviolet exposure. But in every case, especially in the elderly, a search should be made to discover secondary causes such as malabsorption, (including gluten enteropathy and gastrectomy) and interference with the metabolism of vitamin D due to the use of barbiturates and anticonvulsant drugs, liver disease and renal disease.

References

1. DHSS. (1972). A nutritional survey of the elderly. *Report on Public Health and Medical Subjects,* **No. 3.** (London: HMSO)
2. Griffiths, L. L., Brocklehurst, J. C., Scott, D. L., Marks, J. and Blackley, J. (1967). Thiamine and ascorbic acid levels in the elderly. *Geront. Clin.,* 9, 1
3. Taylor, G. F. (1968). A clinical survey of elderly people from a nutritional standpoint. In: A. N. Exton-Smith and D. L. Scott. (eds.). *Vitamins in the Elderly.* (Bristol: John Wright)
4. Brocklehurst, J. C., Griffiths, L. L., Taylor, G. F., Marks, J., Scott, D. L. and Blackley, J. (1968). The clinical features of chronic vitamin deficiency. *Geront. Clin.,* **10,** 309
5. Brin, M. (1964). Erythrocyte as a biopsy tissue for functional evaluation of thiamine status. *J. Am. Med. Assoc.,* **187,** 762
6. Brin, M. (1968). Biochemical methods and findings in U.S. surveys. In: A. N. Exton-Smith and D. L. Scott (eds.). *Vitamins in the Elderly.* (Bristol: John Wright)
7. Philip, G. and Smith, J. F. (1973). Hypothermia and Wernicke's encephalopathy. *Lancet,* ii, 122
8. Hodkinson, H. M. (1973). Mental impairment in the elderly. *J. R. Coll. Physicians.* 7, 305
9. Brin, M., Dibble, M. V., Peel, A., McMullen, E., Bourquin, A. and Chen, N. (1965). Some preliminary findings on the nutritional status of the aged in Onondaga County, New York, *Am. J. Clin. Nutr.,* **17,** 240
10. Thurnham, D. (1972). Riboflavin status of the elderly. (Personal communication)
11. Grant, H. C., Hoffbrand, A. V. and Wells, D. G. (1965). Folate deficiency and neurological disease. *Lancet,* **ii,** 763
12. Strachan, R. W. and Henderson, J. G. (1967). Psychiatric syndromes due to avitaminosis B_{12} with normal blood and marrow. *Q. J. Med.,* **34,** 303
13. Herbert, V. (1967). Biochemical and haematologic lesions in folic acid deficiency. *Am. J. Clin. Nutr.,* **20,** 562
14. Batata, M., Spray, G. H., Bolton, F. G., Higgins, G. and Wollner, L. (1967). Blood and bone marrow changes in elderly patients, with special reference to folic acid, vitamin B_{12}, iron and ascorbic acid. *Br. Med. J.,* **2,** 667
15. Russell, R. I., Williamson, J. M., Goldberg, A. and Wares, E. (1968). Ascorbic acid levels in leukocytes in patients with gastrointestinal haemorrhage. *Lancet,* **ii,** 603
16. Andrews, J., Letcher, M. and Brook, M. (1969). Vitamin C supplementation in the elderly. *Br. Med. J.,* **2,** 416
17. Hyams, D. E. and Ross, E. J. (1963). Scurvy, megaloblastic anaemia, osteoporosis. *Br. J. Clin. Pract.,* **17,** 332
18. Goldberg, A. (1963). The anaemia of scurvy. *Q. J. Med.,* **32,** 51
19. Cox, E. V. (1968). The anaemia of scurvy. *Vit. Horm.,* **26,** 635
20. Stokes, P. L., Melikian, V., Leeming, R. L., Portman-Graham, H., Blair, J. A. and Cooke, W. T. (1975). Folate metabolism in scurvy. *Am. J. Clin. Nutr.,* **28,** 126
21. Walker, A. (1968). Chronic scurvy. *Br. J. Dermatol.,* **80,** 625
22. Kinsman, R. A. and Hood, J. (1971). Some behavioural effects of ascorbic acid deficiency. *Am. J. Clin. Nutr.,* **24,** 455
23. Burr, R. G. and Rajan, K. T. (1972). Leukocyte ascorbic acid and pressure sores in paraplegia. *Br. J. Nutr.,* **28,** 275
24. Irvin, T. I., Chattopadhyay, D. K. and Smythe, A. (1978). Ascorbic acid requirements in postoperative patients. *Surg. Gynecol. Obstet.,* **147,** 49
25. Milne, J. S., Lonergan, M. E., Williamson, J., Moore, F. M. L., McMaster, R. and Percy, N. (1971). Leukocyte ascorbic levels and vitamin C intake in older people. *Br. Med. J.,* **4,** 383
26. Bartley, W., Krebs, H. A. and O'Brien, J. R. P. (1953). Vitamin C requirements of human adults. *MRC Special Report Series,* **No. 280.** (London: HMSO)

27. DHSS. (1969). Recommended intakes of nutrients for the United Kingdom. *Report on Public Health and Medical Subjects*, **No. 120**. (London: HMSO)

28. Windsor, A. C. M. and Williams, C. B. (1970). Urinary hydroxyproline in the elderly with low ascorbic acid levels. *Br. Med. J.*, **i**, 732

29. Fox, R. H., Woodward, P. M., Exton-Smith, A. N., Green, M. F., Donnison, D. V. and Wicks, M. H. (1973). Body temperatures in the elderly; a national study of physiological, social and environmental conditions. *Br. Med. J.*, **i**, 200

30. Aaron, J. E., Gallagher, J. C., Anderson, J., Stasiak. L., Longton, E. B., Nordin, B. E. C. and Nicholson, M. (1974). Frequency of osteomalacia and osteoporosis in fractures of the proximal femur. *Lancet*, **i**, 229

31. Aaron, J. E., Gallagher, J. C. and Nordin, B. E. C. (1974). Seasonal variation of osteomalacia in femoral neck fractures. *Lancet*, **ii**, 84

32. Faccini, J. M., Exton-Smith, A. N. and Boyde, A. (1976). Disorders of bone and fracture of the femoral neck. *Lancet*, **i**, 1089

33. Brown, I. R. F., Bakowska, A. and Millard, P. H. (1976). Vitamin D status of patients with femoral neck fractures. *Age Ageing*, **5**, 127

34. Smith, R. and Dick, M. (1968). Total urinary hydroxyproline excretion after administration of vitamin D to healthy human volunteers and to a patient with osteomalacia. *Lancet*, **i**, 279

35. Bordier, P., Matrajt, H., Hioco, D., Hepner, G. W., Thompson, G. R. and Booth, C. C. (1968). Subclinical vitamin D deficiency following gastric surgery. *Lancet*, **i**, 437

36. Whittle, H., Blair, A., Neale, G., Thalassinos, N., McLaughlin, M., Marsh, M. N., Peters, P. J., Wedzicha, B. and Thompson, G. R. (1969). Intravenous vitamin D in the detection of vitamin D deficiency, *Lancet*, **i**, 747

15
Vitamin deficiencies in disease states

J. KELLEHER

Circulating blood vitamin levels have been assessed in many disease states and in various population groups. The purpose of this chapter is to document the frequency of abnormal blood vitamin levels in different disease states, to discuss the mechanism of these deficiencies and to outline some of the difficulties in their interpretation. Florid clinical vitamin deficiencies such as scurvy and beri-beri though still seen in the UK are now considered extremely rare and, as they are discussed elsewhere in this book (see Exton-Smith, pp. 127–37) they will not be considered further here. This contribution will confine itself to the so-called 'sub-clinical' deficiencies, where the circulating blood vitamin levels are abnormal often in the absence of obvious clinical signs of deficiencies.

MALABSORPTION

Malnutrition is common in malabsorption especially in patients with steator-rhoea and low blood vitamin levels are very frequently observed. Figure 1 shows the proportion of patients with steatorrhoea whom we have observed to have abnormal blood vitamin levels. Abnormalities are frequent and this applies to water-soluble as well as fat-soluble vitamins. The incidence of low folic acid and vitamin B_{12} is similar to many other studies in steatorrhoea and since these two vitamins have been extensively studied in patients with steatorrhoea[1-3] they will not be discussed further here.

Abnormalities of the fat-soluble vitamins are frequent in steatorrhoea as demonstrated in Figure 1, almost 50% of subjects have a low serum vitamin A and over 50% a low serum vitamin E, while 25% have a low serum carotene. These abnormalities are well documented[4-6] and recognized, and subjects with steatorrhoea are usually screened for possible fat-soluble vitamin deficiency and given supplements as required. The incidence of vitamin A and

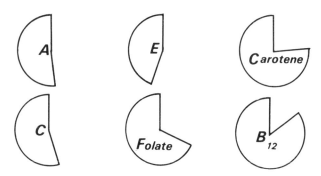

Figure 1 Incidence of low serum vitamin levels in patients with steatorrhoea. The missing segment of each circle represents the proportion of patients with a low level

E deficiency is not determined by the aetiology of the steatorrhoea but the incidence of abnormalities does increase with increasing severity of the steatorrhoea.

Abnormalities of the other fat-soluble vitamins D and K may also occur in steatorrhoea. It is not possible to assess vitamin K status by direct measurement of the vitamin, but prothrombin time measurement and bleeding tendencies both point to a frequent deficiency of this vitamin in various forms of steatorrhoea[7][8] except in pancreatic disease[9] or the blind loop syndrome[10]. Bone disease resulting from vitamin D deficiency is a well-recognized complication in patients with steatorrhoea[11] and more recently with the availability of radioimmunoassay procedures for vitamin D metabolites in serum, it has been demonstrated that a high proportion of patients with fat malabsorption have low serum levels of 25-hydroxycholecalciferol[12]. Deficiency of vitamin D may also result in malabsorption of calcium.

Apart from folic acid and vitamin B_{12}, the other B-complex vitamins have been relatively little studied in patients with malabsorption. Low levels of vitamin B_6 have been reported in various forms of malabsorption[13][14] and while frank clinical deficiency has been rarely described[15] it is evident that subclinical deficiency may be more frequent. There is little evidence of riboflavin deficiency in steatorrhoea though clinical syndromes responding rapidly to riboflavin therapy have been described[16][17]. More extensive use of recently developed biochemical tests for riboflavin deficiency may identify an increased frequency of subclinical riboflavin deficiency. From available evidence it is unlikely that clinical deficiency of vitamin B_1 is common in steatorrhoea though again this vitamin has not been extensively studied in such patients.

Leukocyte ascorbic acid (LAA) levels have not been extensively studied in subjects with steatorrhoea, though this method of assessing tissue levels and thus stores of vitamin C is accepted as being the most useful available[18]. It is perhaps surprising that such a high percentage of subjects with steatorrhoea

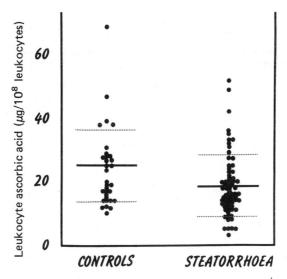

Figure 2 Leukócyte ascorbic acid (LAA) levels in control patients and patients with steatorrhoea. Results are expressed in μg/10⁸ leukocytes, the continuous bar is the mean for each group

(50%) have apparently low tissue stores of vitamin C (Figure 1) though the severity of depletion is not indicated in this figure. Figure 2 shows the individual LAA levels in the subjects with steatorrhoea and in control subjects. The controls were a group of hospital patients without malabsorption. The range of LAA levels in steatorrhoea is extremely variable, from healthy normal to extreme abnormality. Some of these subjects had extremely low values indeed, less than 5 μg/10⁸ WBC, and yet it is emphasized that none of these subjects had clinical evidence of scurvy. There was no correlation between LAA and the severity of steatorrhoea, and the aetiology of the steatorrhoea did not influence the LAA level. As will be shown later, low LAA levels are common in many other disease states.

The causes of vitamin deficiency in steatorrhoea may well be multiple though malabsorption of individual vitamins must be considered of primary importance. This is especially true of the fat-soluble vitamins and malabsorption of vitamins A[6], D[19], E[20] and K[21] has been directly demonstrated. There is also little doubt that as the severity of steatorrhoea increases so also does the malabsorption of the fat-soluble vitamin and there is a reasonably good correlation between the severity of fat and fat-soluble vitamin malabsorption[19] [20]. Bile salts are essential for efficient absorption of fat-soluble vitamins and in diseases accompanied by bile salt deficiency, malabsorption of fat-soluble vitamins is inevitable[22] [23] and may be extremely difficult to correct, even when water miscible preparations of the vitamins are available[24].

Malabsorption of water-soluble vitamins may also occur in steatorrhoea and this is well documented for folic acid and vitamin B$_{12}$. Folic acid malabsorption is common in many forms of steatorrhoea[25], while vitamin B_{12} malabsorption is especially common if the ileum is involved[26].

There is also some evidence that malabsorption of other water-soluble vitamins occurs in steatorrhoea. Thus malabsorption of vitamin B_6 has been demonstrated in coeliac disease[27], while absorption of this vitamin may be normal in other forms of steatorrhoea[27]. Malabsorption of vitamin B_1 has also been demonstrated in coeliac disease but not in Crohn's disease[28]. Perhaps these indicate that when the upper small intestinal mucosa is damaged then malabsorption of water-soluble vitamins occurs but not if the mucosa is normal. Absorption of vitamin C appears to be normal in most forms of steatorrhoea[29].

As well as absorption other factors must also be considered in the aetiology of vitamin deficiencies in steatorrhoea. Williamson *et al.*[30] demonstrated vitamin C deficiency in patients with malabsorption and also after gastric surgery, and speculated that while absorption of the vitamin was probably normal in these subjects as was dietary intake other factors must be involved in producing the deficiency. It has been suggested that increased utilization may exist in such patients[30] [31]. Weight loss is common in patients with malabsorption and in many cases is the commonest presenting feature. Both diminished energy intake and increased energy loss are important causes, the former being perhaps the more important[32] [33]. Diminished energy intake will almost certainly be accompanied by decreased vitamin intake, and a poor dietary intake must also be considered high on the list of factors responsible for vitamin deficiency in malabsorption. It is also worth emphasizing that patients with malabsorption may well have their diets manipulated for thera-

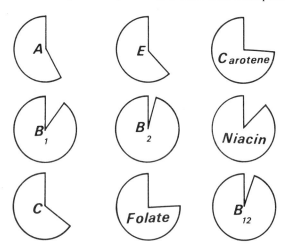

Figure 3 Incidence of low serum vitamin levels in patients with chronic non-alcoholic liver disease. (Adapted from reference 36.) The missing segment of each circle represents the proportion of patients with a low level

peutic purposes which could result in iatrogenic vitamin deficiency unless these deficiencies are recognized and corrected.

Low serum albumin is common in steatorrhoea; over 70% of the patients shown in Figure 1 had a decreased serum albumin. Low circulating proteins may influence the interpretation of circulating vitamin levels. Some vitamins especially fat-soluble ones are dependent on specific carrier proteins for transport in the serum. This is exemplified by vitamin A, low levels of which may be seen in steatorrhoea because adequate levels of the specific retinol-binding protein are not available for transport of normal serum levels[34]. Thus low levels of vitamin A may not necessarily indicate deficiency and may have to be interpreted with caution. The interpretation of serum vitamin A may be further complicated in steatorrhoea by the fact that normal levels do not necessarily exclude deficiency[35].

LIVER DISEASE

For reasons to be discussed later, it is important to distinguish between alcoholic and non-alcoholic liver disease in studying vitamin deficiency. Figure 3 summarizes the frequency of low blood vitamin levels in patients with chronic non-alcoholic liver disease[36]. Low levels of vitamins A and E were common in this study, approximately 40% of patients having low levels, while 25% of subjects had a low serum carotene. Serum folate was low in approximately 25% and leukocyte vitamin C levels were frequently abnormal being low in almost 40% of patients. Only a small proportion of patients had low blood levels of vitamin B_1, B_2, B_{12} or niacin, the latter determined by urinary N-methyl-nicotinamide excretion. However, as with many other blood components the interpretation of blood vitamin levels is complex in liver disease, as will be discussed later.

Other studies of vitamin levels in non-alcoholic liver disease have been performed and the frequency of abnormal results has not always agreed with the findings shown in Figure 3. Thus Rossouw et al.[37] found a much higher incidence of thiamin deficiency and suggested that the difference in incidence between the two series may be due to the fact that their patients were more severely ill. A recent survey of LAA levels in liver disease, while not stating the frequency of abnormalities, found that the mean level in chronic non-alcoholic liver disease was similar to control patients[38]. These discrepancies between different series suggest that it is not possible to extrapolate from one group of liver patients to another and the incidence of vitamin deficiency may vary widely between different centres.

The type of vitamin deficiency found in alcoholic liver disease is often more frequent and severe than in non-alcoholic liver disease and the pattern of vitamin abnormalities found may also be different. Leevy et al.[39-41] in extensive studies on large groups of patients with alcoholic liver disease have shown frequent abnormalities of blood vitamins. Each of 140 patients with alcoholic cirrhosis had low circulating levels of at least two vitamins. In contrast to the data in non-alcoholic liver disease (Figure 3) the most frequent abnormality found by Leevy *et al.* was in the B-complex vitamins; almost 80% of his patients had low serum folate followed by 60% with low vitamin B_6 and B_1, while only a relatively small proportion had low fat-soluble vitamin levels. Low LAA may also be more common in alcoholic than in non-alcoholic liver disease[38], though in other series the incidence of low LAA has not been so great.

The causes of vitamin deficiency in liver disease are many and complex and have recently been reviewed[42 43]. Dietary intake is obviously of prime importance especially in the alcoholic patient, and a poor dietary intake may well be the most important cause of vitamin deficiency in liver disease. Diseases of the liver are often accompanied by anorexia, nausea, vomiting and food intolerance[43] and it is therefore not surprising that dietary intake is an important factor in determining nutritional status in such patients. Inadequate diet is much more frequent in alcoholic as compared with non-alcoholic liver disease; over 60% of alcoholics were demonstrated as having a poor intake[40], a much higher percentage than in a non-alcoholic group[36].

Malabsorption must also be considered as a causative factor of vitamin deficiencies in liver disease. Steatorrhoea is present in up to 50% of patients with liver disease[44], regardless of aetiology. In the patients whose vitamin levels are shown in Figure 3, almost 40% had steatorrhoea, though it was usually not very severe. The presence of steatorrhoea may well directly influence the fat-soluble vitamin status of such patients, but is unlikely to be a major determinant of water-soluble vitamin status. In alcoholic liver disease vitamin malabsorption is important as alcohol may directly affect the absorption of some water-soluble vitamins[45-47]. It has recently been demonstrated that cholestyramine which reduces intraluminal bile salt concentration may, as well as its well recognized effects on the absorption of fat soluble vitamins[48 49], also results in reduced LAA levels[38].

The liver is the major storage organ for most vitamins, and it has been reported that hepatic levels of folic acid, riboflavin, pyridoxine, vitamin B_{12} and vitamin A[39 41 50] are often decreased in chronic liver disease. More recently, a significant correlation was demonstrated between LAA and liver ascorbic acid levels in chronic liver disease[38]. The cause of the decreased storage of vitamins in liver disease could be due either to decreased uptake of absorbed vitamin or increased release from a damaged liver.

As well as storage the liver is also an important site of metabolism of vitamins to their active metabolites and is also the site of synthesis of proteins required for vitamin transport in serum. Pyridoxal-5-phosphate, the active coenzyme form of vitamin B_6, has been shown to be low in a large proportion of alcoholic[51] and non-alcoholic[52] liver disease patients. When intravenous vitamin B_6 is administered to these patients the resultant rise in serum pyridoxal-5-phosphate is much less than in controls, and following intravenous pyridoxal-5-phosphate its rate of degradation is far greater in liver disease patients compared with controls. It has been suggested that increased degradation of pyridoxal-5-phosphate may be an important cause of vitamin B_6 deficiency in liver disease subjects. Abnormalities of thiamin metabolism may also occur in liver disease as very large doses of supplemental thiamin are required to correct the abnormal erythrocyte transketolase activity seen in such subjects[37].

Various vitamins are transported in serum by proteins synthesized in the liver. Vitamin A is one example and is transported by a specific protein synthesized in the liver, retinol-binding protein. Levels of retinol-binding protein are decreased in chronic liver disease[34 50] and may influence the interpretation of plasma vitamin A levels. Dark adaptation may be impaired in some patients with chronic liver disease and may not respond completely to vitamin A therapy[53 54] though a deficiency of zinc may also be causative[54].

OTHER HOSPITAL GROUPS AND DISEASES

Plasma vitamin levels have been assessed in many other groups of hospital patients. The literature on vitamin levels in the elderly is extensive and variable, the incidence of deficiency varying widely depending on the particular elderly group studied. Several recent reviews on this subject have been published[55 56]. Figure 4 summarizes the incidence of abnormal serum vitamin

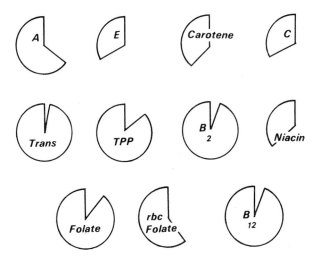

Figure 4 Incidence of low serum vitamin levels in acute geriatric admissions. (Adapted from reference 58.) The missing segment of each circle represents the proportion of patients with a low level

levels found in one group of elderly subjects admitted to the geriatric unit of this hospital[57 58]. Frequent abnormalities were observed, as were multiple vitamin deficiencies in many patients. The most frequently low results were LAA and vitamin E, almost 70% of patients having low levels followed by vitamin A and erythrocyte folic acid which were abnormal in almost 40% of these patients. Niacin status as assessed by urinary N-methyl-nicotinamide excretion was low in over 60% of patients. Less frequent abnormalities were found for thiamin, riboflavin and vitamin B_{12}. In this particular series of patients the most likely cause of deficiency was an inadequate diet though, of course, other causes of vitamin deficiency may be of importance in elderly patients. It is emphasized that this study was performed in a particular, selected group of elderly individuals and while other studies have demonstrated a similar frequency of abnormalities for particular vitamins[59-61] these reports have usually concerned hospitalized or institutionalized groups. The selection of the elderly group for study appears to be the important factor in determining the frequency of abnormalities which various studies have reported.

Nutritional depletion is a common hallmark of patients with cancer[62] and there are many factors in such patients such as anorexia, mechanical obstruction, chemotherapy and radiotherapy, which may contribute to the development of vitamin deficiency. Calman[63] recently reported blood vitamin levels in 120 cancer patients and reported frequent abnormalities. Folic acid, LAA and thiamin were abnormal in a very high proportion of patients and vitamin A was less frequently abnormal. Vitamin B_{12} was normal in all patients again perhaps reflecting the large stores of this vitamin available in the body. The high proportion of cancer patients with low circulatory levels of blood vitamins suggests that deficiency may be frequent, and this may persist for some vitamins even when multivitamin supplements are being administered[64]. Calman[63] also points out that because of the role of ascorbic acid in the development of cancer and its role in drug metabolism[65] this vitamin may have special significance in such patients. Vitamin A status may modify the response rate to chemotherapy in cancer patients and its depletion may depress the immune response.

Table 1 Percentage of patients in different hospitalized groups with low LAA levels

	% Abnormal	Reference
Chronic non-alcoholic liver disease	35	36
Alcoholic liver disease	57	38
Malabsorption	43	This study
Cancer	71	63
Surgical patients	25	66
Elderly	70	58

There is little doubt that subclinical vitamin deficiency is present in certain groups of hospital patients, and the proportion of patients with abnormalities of circulating vitamins is sometimes high. Table 1 gives examples of the frequency of low LAA levels which have been reported. This table is in no way representative of the ascorbic acid status of all vulnerable patient groups but serves to demonstrate that if such abnormalities are carefully looked for in specific groups, a high proportion of patients may well be shown to have sub-clinical deficiency. This applies to other vitamins as well as ascorbic acid and raises the question as to whether vitamin levels should be more frequently measured, especially in hospitalized and institutionalized groups.

References

1. Badenoch, J. (1960). Steatorrhoea in the adult. *Br. Med. J.*, **2**, 879 and 963
2. Klipstein, F. A. (1966). Folate deficiency secondary to diseases of the intestinal tract. *Bull. N.Y. Acad. Sci.*, **42**, 638
3. Cooke, W. T. (1958). Adult coeliac disease and other disorders associated with steatorrhoea. *Br. Med. J.*, **2**, 261
4. Binder, H. J., Herting, D. C. and Hurst, V. (1965). Tocopherol deficiency in man. *N. Engl. J. Med.*, **273**, 1289
5. Goransson, G., Norden, A. and Akesson, B. (1973). Low plasma tocopherol levels in patients with gastrointestinal disease. *Scand. J. Gastroenterol.*, **8**, 21
6. Kahan, J. (1970). The vitamin A absorption test. II. Studies on children and adults with disorders in the alimentary tract. *Scand. J. Gastroenterol.*, **5**, 1
7. Bossak, E. T., Wang, C. I. and Aldersberg, D. (1957). Clinical aspects of the malabsorption syndrome (idiopathic sprue). *J. Mt. Sinai Hospital*, **24**, 261
8. Green, P. A. and Wollaeger, E. E. (1960). The clinical behaviour of sprue in the United States. *Gastroenterology*, **38**, 399
9. Evans, W. B. and Wollaeger, E. E. (1966). Incidence and severity of nutritional deficiency states in chronic exocrine pancreatic insufficiency: comparison with non-tropical sprue. *Am. J. Digest. Dis.*, **11**, 594
10. Gracey, M. (1971). Intestinal absorption in the 'contaminated small-bowel syndrome'. *Gut*, **12**, 403
11. Nordin, B. E. C. (1961). Effect of malabsorption syndrome on calcium metabolism. *Proc. R. Soc. Med.*, **54**, 497
12. Schoen, M. S., Lindenbaum, J., Roginsky, M. S. and Holt, P. R. (1978). Significance of serum level of 25-hydroxycholecalciferol in gastrointestinal disease. *Digest. Dis.*, **23**, 137
13. Anderson, B. B., Peart, M. B. and Fulford-Jones, C. E. (1970). The measurement of serum pyridoxal by a microbiological assay using lactobacillus casei. *J. Clin. Pathol.*, **23**, 232
14. Baker, H. and Sobotka, H. (1962). Microbiological assay method of vitamins. *Adv. Clin. Chem.*, **5**, 173
15. Dawson, A. M., Holdsworth, C. D. and Pitcher, C. S. (1964). Sideroblastic anaemia in adult–coeliac disease. *Gut*, **5**, 304
16. Neale, G., Antcliff, A. C., Welbourn, R. B., Mollin, D. L. and Booth, C. C. (1967). Protein malnutrition after partial gastrectomy. *Q. J. Med.*, **36**, 469
17. Lai, C. and Ransome, G. A. (1970). Burning-feet syndrome. Case due to malabsorption and responding to riboflavine. *Br. Med. J.*, **2**, 151
18. Sauberlich, H. E., Dowdy, R. P. and Skala, J. H. (1973). Laboratory tests for the assessment of nutritional status. *Crit. Rev. Clin. Lab. Sci.*, **4**, 227
19. Thompson, G. R., Lewis, B. and Booth, C. C. (1966). Absorption of vitamin D_3-^{3}H in control subjects and patients with intestinal malabsorption. *J. Clin. Invest.*, **45**, 94
20. Kelleher, J. and Losowsky, M. S. (1970). The absorption of α-tocopherol in man. *Br. J. Nutr.*, **24**, 1033
21. Shearer, M. J., Mallinson, C. N., Webster, G. R. and Barkhan, P. (1970). Absorption of tritiated vitamin K_1 in patients with fat malabsorption. *Gut*, **11**, 1063
22. Thompson, G. R. (1971). Absorption of fat-soluble vitamins and sterols. In: A. M. Dawson

(ed.). *Intestinal Absorption and its Derangements,* pp 85–89. (BMA House, London: *J. Clin. Pathol.*)

23. Forsgren, L. (1969). Studies on the intestinal absorption of labelled fat-soluble vitamins (A, D, E and K) via the thoracic-duct lymph in the absence of bile in man. *Acta. Chir. Scand.*, **399** (Suppl), 5

24. Harries, J. T. and Muller, D. P. R. (1971). Absorption of vitamin E in children with biliary obstruction. *Gut,* **12**, 579

25. Bernstein, L. H., Gutstein, S., Weiner, S. and Efron, G. (1970). The absorption and malabsorption of folic acid and its polyglutamates. *Am. J. Med.*, **48**, 570

26. Toskes, P. P. and Deren, J. J. (1973). Vitamin B_{12} absorption and malabsorption. *Gastroenterology,* **65**, 662

27. Brain, M. C. and Booth, C. C. (1964). The absorption of tritium-labelled pyridoxine HCl in control subjects and in patients with intestinal malabsorption. *Gut,* **5**, 241

28. Thompson, A. D. (1966). The absorption of radioactive sulphur-labelled thiamine hydrochloride in control subjects and in patients with intestinal malabsorption. *Clin. Sci.*, **31**, 167

29. Stewart, J. S. and Booth, C. C. (1964). Ascorbic acid absorption in malabsorption. *Clin. Sci.*, **27**, 15

30. Williamson, J. M., Goldberg, A. and Moore, F. M. L. (1967). Leukocyte ascorbic acid levels in patients with malabsorption or previous gastric surgery. *Br. Med. J.*, **2**, 23

31. *Nutrition Reviews* (1967). Malabsorption, gastric surgery and ascorbic acid. *Nutr. Rev.*, **25**, 237

32. Comfort, M. W., Wollaeger, E. E., Taylor, A. B. and Power, M. H. (1953). Non-tropical sprue: Observations on absorption and metabolism. *Gastroenterology,* **23**, 155

33. Culver, P. J. (1962). Postvagotomy and gastrectomy — nutrition and steatorrhoea. *Ann. N. Y. Acad. Sci.*, **99**, 213

34. Vahlquest, A., Sjolund, K., Norden, A., Peterson, P. A., Stigmar, G. and Johansson, B. (1978). Plasma vitamin A transport and visual dark adaptation in diseases of the intestine and liver. *Scand. J. Clin. Lab. Invest.*, **38**, 301

35. Russell, R. M., Smith, V. C., Multack, R., Krill, A. E. and Rosenberg, I. H. (1973). Dark-adaptation testing for diagnosis of sub-clinical vitamin-A deficiency and evaluation of therapy. *Lancet,* **ii**, 1161

36. Morgan, A. G., Kelleher, J., Walker, B. E. and Losowsky, M. S. (1976). Nutrition in cryptogenic cirrhosis and chronic aggressive hepatitis. *Gut,* **17**, 113

37. Rossouw, J. E., Cabadarios, D., Krasner, N., Davis, M. and Williams, R. (1978). Red blood cell transketolase activity and the effect of thiamine supplementation in patients with chronic liver disease. *Scand. J. Gastroenterol.*, **13**, 133

38. Beattie, A. D. and Sherlock, S. (1976). Ascorbic acid deficiency in liver disease. *Gut,* **17**, 571

39. Baker, H., Frank, D., Ziffer, H., Goldfarb, S., Leevy, C. M. and Sobotka, H. (1964). Effect of hepatic disease on liver B-complex vitamin titers. *Am. J. Clin. Nutr.*, **14**, 1

40. Leevy, C. M., Baker, H., Tenhove, W., Frank, O. and Cherrick, G. R. (1965). B-complex vitamins in liver disease of the alcoholic. *Am. J. Clin. Nutr.*, **16**, 339

41. Leevy, C. M., Thompson, A. and Baker, H. (1970). Vitamins and liver injury. *Am. J. Clin. Nutr.*, **23**, 493

42. Mezey, E. (1978). Liver disease and nutrition. *Gastroenterology,* **74**, 770

43. McIntyre, N. and Bateman, C. (1978). Disorders of liver and gall bladder. In: J. W. T. Dickerson and H. A. Lee (ed.). *Nutrition in the Clinical Management of Disease* pp. 192–209. (London: Edward Arnold Publishers Ltd.)

44. Losowsky, M. S. and Walker, B. E. (1969). Liver disease and malabsorption. *Gastroenterology,* **56**, 589

45. Mezey, E. (1975). Intestinal function in chronic alcoholism. *Ann. N. Y. Acad. Sci.*, **252**, 215

46. Halsted, C. H., Griggs, R. C. and Harris, J. W. (1967). The effect of alcoholism on the absorption of folic acid (H^3-PGA) evaluated by plasma levels and urine excretion. *J. Lab. Clin. Med.*, **69**, 116

47. Thomson, A. D., Baker, H. and Leevy, C. M. (1970). Patterns of ^{35}S-thiamine hydrochloride absorption in the malnourished alcoholic patient. *J. Lab. Clin. Med.*, **76**, 34

48. Thompson, W. G. and Thompson, G. R. (1969). Effect of cholestyramine on the absorption of vitamin D_3 and calcium. *Gut,* **10**, 717

49. Davies, T., Kelleher, J., Smith, C. L., Walker, B. E. and Losowsky, M. S. (1972). Effect of

therapeutic measures which alter fat absorption on the absorption of α-tocopherol in the rat. *J. Lab. Clin. Med.*, **79**, 824

50. Smith, F. R. and Goodman, De Witt, S. (1971). The effects of diseases of the liver, thyroid and kidneys on the transport of vitamin A in human plasma. *J. Clin. Invest.*, **50**, 2426
51. Mitchell, D., Wagner, C., Stone, W. J., Wilkinson, G. R. and Schenker, S. (1976). Abnormal regulation of plasma pyridoxal-5¹-phosphate in patients with liver disease. *Gastroenterology*, **70**, 988
52. Labadarios, D., Rossouw, J. E., McConnell, J. B., Davis, M. and Williams, R. (1977). Vitamin B6 deficiency in chronic liver disease — evidence for increased degradation of pyridoxal-5¹-phosphate. *Gut*, **18**, 23
53. Patek, A. J. and Haig, C. (1939). The occurrence of abnormal dark adaptation and its relation to vitamin A metabolism in patients with cirrhosis of the liver. *J. Clin. Invest.*, **18**, 609
54. Morrison, S. A., Russell, R. M., Carney, E. A. and Oaks, E. V. (1978). Zinc deficiency: a cause of abnormal dark adaptation in cirrhotics. *Am. J. Clin. Nutr.*, **31**, 276
55. Exton-Smith, A. N. (1978). Nutrition in the elderly. In: J. W. T. Dickerson and H. A. Lee (eds.). *Nutrition in the Clinical Management of Disease*, pp. 72–104. (London: Edward Arnold Publishers Ltd.)
56. Hyams, D. E. (1973). Nutrition in the elderly. *Modern Geriatrics*, **3**, 352
57. Morgan, A. G., Kelleher, J., Walker, B. E., Losowsky, M. S., Droller, H. and Middleton, R. S. W. (1973). A nutritional survey in the elderly: Haematological aspects. *Int. J. Vit. Nutr. Res.*, **43**, 461
58. Morgan, A. G., Kelleher, J., Walker, B. E., Losowsky, M. S., Droller, H. and Middleton, R. S. W. (1975). A nutritional survey in the elderly: Blood and urine vitamin levels. *Int. J. Vit. Nutr. Res.*, **45**, 448
59. Milne, J. S., Lonergan, M. E., Williamson, J., Moore, F. M. L., McMaster, R. and Percy, N. (1971). Leukocyte ascorbic acid levels and vitamin C intake in older people. *Br. Med. J.*, **4**, 383
60. Griffiths, L. L., Brocklehurst, J. C., Scott, D. L., Marks, J. and Blackley, J. (1967). Thiamine and ascorbic acid levels in the elderly. *Geront. Clin.*, **9**, 1
61. Read, A. E., Gough, K. R., Pardoe, J. L. and Nicholas, A. (1965). Nutritional studies on the entrants to an old people's home with particular reference to folic-acid deficiency. *Br. Med. J.*, **4**, 843
62. Editorial. (1978). Nutrition and the patient with cancer. *Br. Med. J.*, **3**, 84
63. Calman, K. C. (1978). Nutritional support in malignant disease. *Proc. Nutr. Soc.*, **37**, 87
64. Basu, T. K., Dickerson, J. W. T., Raven, R. W. and Williams, D. C. (1974). The thiamine status of patients with cancer as determined by the red cell transketolase activity. *Int. J. Vit. Nutr. Res.*, **44**, 53
65. Zannoni, V. G., Flynn, E. J. and Lynch, M. (1972). Ascorbic acid and drug metabolism. *Biochem. Pharmacol.*, **21**, 1377
66. Hill, G. L., Blackett, R. L., Pickford, I., Burkinshaw, L., Young, G. A., Warren, J. V., Schorah, C. J. and Morgan, D. B. (1977). Malnutrition in surgical patients: an unrecognised problem. *Lancet*, i, 689

16
Progress in the prevention and treatment of nutritional rickets

G. C. ARNEIL

Two years ago at a previous symposium a simple classification of nutritional rickets was used[1] which has been updated as follows:-

Congenital rickets
Preterm rickets
Infantile rickets
Toddler rickets
Asian adolescent rickets
Calcium deficiency rickets

The recent progress in each of these categories, in prevention and in treatment, where relevant, will be considered in turn.

CONGENITAL RICKETS

This is the most florid example of foetal hypovitaminosis D and in consequence one which has attracted a great deal of attention relative to the number of known cases. The condition was described in 1930 by Maxwell and Turnbull[2] and in Asians in Britain by Ford et al.[3] in 1973 (two cases) and by Moncrieff and Fadahunsi[4] in 1974.

The complex situation pertaining to the foetus *in utero* is related not only to maternal vitamin D intake and metabolism but also to placental function and to vitamin D metabolism in the foetus. Calcium intake and metabolism in the mother and foetus and probably parathyroid hormone activity in both also play a part. Classical congenital rickets usually presents as neonatal hypocalcaemic seizures. It is identified radiologically as being due to vitamin D deficiency and the diagnosis is confirmed later by measuring serum calcium, phosphorus, alkaline phosphatase and 25-hydroxycholecalciferol (25-HCC) levels. Craniotabes may be detected. Treatment given by Ford *et al.*[3], and Moncrieff and Fadahunsi[4] tended to be orthodox with calcium being given initially followed by vitamin D therapy.

In a fascinating study in Meerut (India) Teotia *et al.*[5] have described their findings in 500 pregnant women. The recommended intake of vitamin D for pregnant women is 400 IU daily save in India where the Indian Council of Medical Research advises 200 IU daily[6]. The dietary intake of 500 women from Meerut was estimated to average 30 IU daily. Similarly, calcium intake for pregnant women is usually recommended at 1.0–1.2 g daily but in Meerut the average intake was 300 mg daily.[5]

From these 500 pregnancies Teotia *et al.*[5] studied 110 babies, two of whom had congenital rickets and two hypocalcaemia. They report one case in fascinating detail. During the experimental period this baby was totally breast fed and protected from sunshine and skyshine while metabolic studies proceeded. Initially the serum 25-HCC level of the mother and that of the baby were both undetectable. A single intravenous dose of vitamin D_3 healed the mother's osteomalacia but whilst maternal serum 25-HCC levels rose somewhat the level in the infant did not, nor did healing of rickets occur. A second dose of vitamin D intravenously three months later produced healing of rickets in the baby and in both mother and child the serum 25-HCC levels returned to normal. The calcium levels in her breast milk remained satisfactory throughout despite osteomalacia.

These findings raise the fascinating question as to whether in lesser degrees of osteomalacia in Asian women the breast milk levels of vitamin D and/or vitamin D sulphate are inadequate for her nursling. It seems that the foetus has first call on maternal calcium and perhaps vitamin D metabolites *in utero* but that from breast milk only calcium for the nursling receives priority.

If severe florid rickets can develop *in utero* then it seems certain that lesser degrees must be occurring more frequently when osteomalacic mothers are pregnant. Does this affect foetal growth? The reported cases have tended to be small (Teotia *et al.*, 2.2 kg at term,[5] Ford *et al.*, (a) 3.0 kg at term (b), 2.4 kg at 36 weeks[3], Moncrieff and Fadahunsi[4], 2.75 kg at term). Clearly a full metabolic study of vitamin D and calcium metabolism, of parathyroid hormone activity and of foetal growth, osteomalacia and rickets, together with a study of calcium and vitamin D metabolism in the nursling of osteomalacic mothers is overdue. Hypovitaminosis D *in utero* may be uncommon and confied to Asians in Britain but now that it has been recognized in 3.6% of babies studied in Meerut, hypovitaminosis D is likely to be uncovered in other areas.

Cockburn *et al.*[7] have completed a fascinating study on maternal vitamin D intake and calcium metabolism in mothers and newborn infants. They measured plasma calcium, phosphorus, magnesium and total protein levels at intervals during pregnancy and at delivery both in women given supplements of vitamin D_2 from about the 12th week of gestation and in a control group receiving a placebo. Umbilical venous blood and sixth day capillary blood values were measured in several hundred infants. 25-HCC levels were much higher in the maternal, umbilical and infant samples of the vitamin D fortified group. The mother at 24 weeks and the six day infant had significantly higher plasma calcium levels in the vitamin D fortified group making neonatal tetany less likely. There is an urgent need to repeat such studies on Asian women and to ascertain the 25-HCC levels in cord blood as well as breast fed babies. Cockburn et al.[7] noted that breast fed babies of white Scottish mothers had a lower level of 25-HCC than those artificially fed. The Teotias' work[5] suggests that levels may be even lower in young Asian nurslings of osteomalacic mothers. Clearly we are moving from an era of reporting relatively rare florid cases of congenital rickets to a more studied analysis of the problems, including the damage caused to the enamel of teeth by neonatal hypocalcaemia.

PRETERM RICKETS

This condition, formerly called rickets of prematurity, is now terminologically brought into line with modern thought which regards real gestational age as opposed to bodyweight as the desirable definitive parameter. The increased vitamin D requirement of the preterm newborn is so well recognized that in efficient health care areas preterm rickets should no longer occur. Two new aspects have required consideration. Firstly, the need to improve and standardize the care and storage of breast milk in banks raises the possibility of (a) losing vitamin D or vitamin D sulphate in the process and (b) fortifying this aliment. Secondly, the vitamin D and vitamin D sulphate content of breast milk from lactating vitamin D deficient mothers is pertinent as mentioned above.

Preterm rickets and craniotabes, the quasi-physiological consequences of very rapid linear growth in the preterm infant, contrasts with the rapid weight gain of the low birth-weight dysmature ('light-for-dates') term infant whose main problem lies in hypoglycaemia.

A fascinating study in prevention was carried out by Fleischman et al.[8] on 30 preterm babies. Eight were given 25-HCC in dosage of 2 μg/kg/day orally for five days beginning 12 h after birth, while the remaining 22 children acted as controls. Eleven of the 22 controls developed hypocalcaemia, ionized calcium being less than 0.75 mmol/l and total serum calcium level less than 1.75 mmol/l on day 2. Only one of the eight treated infants developed hypocalcaemia. The concentrations of serum 25-HCC, total calcium and ionized calcium in the eight treated patients were considerably greater than in the controls during the first five days of life. No untoward effects were observed.

Once more we are moving towards an era when the use of vitamin D therapy, as 25-HCC or other metabolite, must be taken seriously as a method of avoiding neonatal hypocalcaemia with its very unpleasant potential consequences of neonatal tetany and dental dysplasia. Since immediate

153

postnatal hepatic function is so variable in the preterm baby this may well be a situation where 25-HCC treatment is specifically indicated.

It is interesting that the concentration of vitamin D sulphate in early breast milk is reported to be 20 μg/l as opposed to 8 μg/l in later breast milk.

INFANTILE RICKETS

This was the formerly very common and classical result of the use of unfortified substitutes for breast milk from soon after birth. The earliest sign is craniotabes followed within a few months by costo-chondral beading, cupping and fraying of the epiphyseal ends of the metaphyses of the long bones and bowing of the legs as excessive parathyroid hormone activity complicates earlier subclinical rickets. Abundant in Britain until 1944, infantile rickets was almost entirely prevented by the introduction of National Dried Milk (NDM). The demerits of so-called 'idiopathic hypercalcaemia' and of the high solute load pertaining to NDM should not be allowed to obscure the great benefits it brought not only in despatching infantile rickets but also in largely preventing severe infantile gastroenteritis, until then a frequent cause of infantile malnutrition and death.

Today there are two main concerns in prevention. The first is how to curb and to educate the uninformed and/or perverse mothers who still feed the young baby liquid 'doorstep milk'(unsterile, unfortified and high solute) from the early days of life. Health education must not ease up on emphasizing the hazards of this practice, the unique merits of breast milk, and the advantages of low-solute baby milks. Secondly there is still a tendency to stop feeding 'babymilks' far too early, at 3-4 months of age. For the sake of ensuring adequate calcium, iron and vitamin intake and relative safety from enteral disorders it is my conviction that babies should desirably remain on fortified 'babymilks' throughout the first year of life. There is nothing to be gained by a switch to 'doorstep milk' and something to be lost. The prevention of infantile rickets therefore depends on continuing health education in schools, of women during pregnancy and the early months of the life of *each* newborn baby remembering that as parity increases so does the carelessness of most mothers and their belief that they 'know it all'. Treatment, where necessary, has not changed.

The division between infantile and toddler rickets is much sharper today than it was 50 years ago when inadequate vitamin D intake began soon after birth and persisted for many years as Glisson so well described centuries ago.

TODDLER RICKETS

Toddler rickets affects mainly those aged nine months to three years who have received an adequate intake of vitamin D from fortified babymilks for the first 4-6 months, but whose intake falls thereafter to a level that is always below 100 IU daily and often below 50 IU daily[9]. This early switch to doorstep milk with no vitamin supplements being given to a considerable proportion of babies led to the considerable return of toddler rickets in Glasgow in the 1960s[9] when subclinical bony changes were first demonstrated and subsequently confirmed[10][11]. Toddler rickets may be severe enough to produce pelvic contracture and is typified by bow-legs (genu varum) in babies already walking.

Sadly as Pakistani, Indian and West Indian mothers have become established in Britain without adequate instruction in health and English being provided they have tended to copy their Anglo-Saxon and Celtic neighbours and swung from natural to unnatural feeding for their babies. Inevitably, then, toddler rickets is appearing in some babies in these ethnic groups. Luckily the Chinese immigrants have so far had more sense and continue with long-term breast feeding[12] and, paradoxically, their problem, in Glasgow at least, is reluctance to supplement breast feeding with mixed feeding until the baby is more than one year old, so that protein and energy intakes are relatively low in the second six months of life.

Health education plus free and easy availability of Vitavel (easier to use and much simpler than Government vitamin tablets with its confusing range of doses in the mini-bottle with micro-instructions) seem to be the best policy to avoid toddler rickets in underprivileged families. The best place for such education is in the home, in the mother's own language, and health visitors are an important key to optimal results. Fortunately West Africans and Chinese in Glasgow have relatively few language problems.

This form of rickets can be contained relatively easily if health services are efficient and it probably occurs in less than 0.5% of Glasgow children at present. Treatment when required remains supplementary vitamin D orally or Stoss therapy followed by advice on including foods rich in vitamin D in the diet or more simply the use of a 'tonic' such as Vitavel (5 ml daily) until school entry. Slum clearances have also played a large part in making outdoor play areas available in the centre of Glasgow, now a smokeless zone. Fortunately toddlers' legs have far to grow and even gross bowing of the legs is almost entirely self-correcting provided adequate dietary and supplemental intake of vitamin D is maintained.

ASIAN ADOLESCENT RICKETS

The situation with regard to adolescent rickets in Britain in 1978 remains unsatisfactory. Despite a wealth of clinical data on this subject since 1962[1][3][4][10-14] and the demonstration of the preventive and curative aspects of vitamin D supplements and of vitamin D fortified chappati flour[15] there has been a failure of the Health Services to take adequate and decisive action. There have been unnecessary delays rather than lack of information or of health personnel.

Two years ago Dr Dunnigan and I in separate papers[1][14] drew attention to the salient points in relation to Asian rickets which merit re-emphasis.

1. Nutritional rickets in British children aged six years or more is virtually all in Pakistanis and Indians, usually Muslims.
2. Such immigrants cling to their religion, dress, language and dietary habits.
3. The diet of many such immigrants contains a large proportion of chappatis and other bread[16] made from whole flour.
4. Several members of the same family may be affected making the possibility of genetically determined factors in intermediary vitamin D metabolism being involved a real possibility.
5. In many immigrant families mothers do not understand or speak English, fathers have a limited vocabulary and only school age children become fluent in English. Food is bought in shops where Urdu or Hindi is spoken.
6. In spite of pain in limbs, awkward and deteriorating mobility and marked knock-knees (genu valgum) the condition is often overlooked by teachers, nurses and doctors alike. Lower thighs and legs must be exposed, the subject stood with knees *just apart,* and feet pointing straight forward. Modesty and purdah tends to veil the deformed limbs of pubertal Asian females more than those of their Scottish counterparts but there can be no excuse for the gross deformities which have been overlooked by school teachers, school health services and some family practitioners.
7. A series of trials has shown that giving extra vitamin D (300–400 IU daily) to such children will prevent and heal such rickets. As long ago as 1976 it was shown that fortifying chappati flour was effective[14 15].
8. Although supplementary vitamin D heals rickets at puberty it does not correct the deformity and since little growth and much weight bearing is involved permanent and progressive joint damage is inevitable. Osteotomy (breaking and re-aligning) of femora and sometimes tibiae as well is involved. In Glasgow between 1970 and 1975 ten children and adolescents required such major surgery and more have been operated on since.
9. As previously discussed continuation of osteomalacia in pregnant young women may lead to fœtal hypovitaminosis D which may result in fœtal and congenital rickets and severe neonatal tetany may result. Furthermore untreated osteomalacic mothers may not provide adequate vitamin D or vitamin D sulphate in breast milk for young babies[5 17].

Although largely confined to cities and focal within these cities a real problem exists. A study in Bradford indicated that 2.5% Asian children may be admitted to hospital with rickets and in Glasgow the figure is nearer 5% of those aged 5–16 years. This takes no heed of the proportion of undiagnosed cases and those treated as outpatients, Goel et al.[11] found 5% with florid rickets and a further 6% with radiological rickets making 11% in all.

The problem in prevention is how to make sure that *all* vulnerable Pakistani and Indian babies, children and adolescents (and pregnant women) avoid relative vitamin D deficiency. In the city of Glasgow facts are easily available thanks to Dr Goel[16]. There are in Glasgow approximately 12000 Asians from India and Pakistan of whom 3500 are children. These include 7500 Muslims, 4300 Sikhs and various small sects. Whereas 89% of such children receive vitamin D supplements in the first year of life only 44% continue for 6–10 years and 27% from ten years onwards. These figures would be much lower were it not for the efforts of health visitors and medical officers in infant clinics following the advice of the various publications produced by Glasgow by local research groups[1 3 9–14 16 18]. Medication of older children is by Vitavel (5 ml daily = 300 IU) or Government vitamin tablets and supplementation will be required as long as high intake of chappatis and high extract flour intake continues. Goel[16] has shown that of 95 Asian children aged six years or more 91 (95%) received chappati flour, 78 consumed 56 g daily or more, 21 ate 110 g daily or more and 15 consumed 170 g daily or more. No Asian child was taking only margarine but 50% took only butter.

Prevention means either teaching English and reading to immigrant women, providing health education and producing a change in dietary habits — a hopeless prospect in the short run — or introducing supplemental vitamin D for all Asian children and adolescents knowingly or unknowingly.

One positive suggestion has been the addition of vitamin D to chappati flour[14 15]. This has some merit but involves complicated legislation; it means that all chappati-eating Asians (70% of whom are not children) will get extra vitamin D in considerable amounts and many children will receive vitamin D supplements in addition. Furthermore the wholemeal flour used for chappatis is used for other purposes and doubtless will be used elsewhere. Although attractive superficially this suggestion does not seem a good solution. What is very disturbing are the years of indecision as to whether or not to fortify chappati flour. This has tended to delay positive action, and prolonged the period during which a considerable proportion of Asian adolescents have been acquiring an unnecessary crippling nutritional deficiency leading in some to permanent deformity requiring major surgical interference.

Gardee[18] reported from information supplied by the Research and Intelligence Unit of the Scottish Home and Health Department and by Ford and Dunnigan that the number of schoolchildren aged 5–16 years admitted to Glasgow hospitals in the period 1970–75 was 66 amounting to 4.9% of the total number of such children in Glasgow (2650 in 1975). Such figures tend to be spurious because more cases are revealed in years when surveys are conducted and fewer detected in other years. Furthermore these figures take no account of outpatients or of those treated by family doctors.

A second possible approach is free supplementary vitamin D for Asian children. Lacking effective advice from central health services, the Greater

Glasgow Health Board Area in November 1978 decided to 'go it alone' and try to prevent further crippling disease in its immigrant children. The programme to be applied from now on is as follows:-

1. Vitamin D supplements (400 IU daily) will be available free of charge to all vulnerable children aged 0–18 years as of right and without means test. This means approximately 300000 children are potentially eligible.

2. The supplement will be as drops, syrup or tablets to suit the age concerned and will be made available via clinics and taken by health visitors to the homes. Schools will be involved if practicable. Since virtually all Asians in Glasgow attend only 15% of schools and if suitable distributive techniques can be achieved this would be the best solution.

3. The Health Education Department and the Community Relations Committee will initiate and continue a health education programme in relevant schools, cinemas and community centres to increase awareness in Pakistanis and Indians of the problem of Asian rickets and how to prevent, identify and cure it. The cartoon film 'In Place of the Sun' and slides from it will be used.

4. An in-service training programme is being arranged for family doctors, medical officers, obstetricians, health visitors, school nurses and antenatal clinical staff in each district of the city.

5. Teachers in schools will annually impress the need for supplementary vitamin D (particularly for Asians) as winter approaches. They will also be urged to inspect the knees and legs of Asians at physical training classes and watch for walking problems, and for knock-knees.

These empirical moves are aimed at countering the rachitogenic tendency known to exist in the community and to reduce the number of children requiring osteotomy in the future, to reduce osteomalacia in pregnant women and hypovitaminosis in the fœtus with possible hazard then and to the newborn baby. One cannot be happy that extra vitamin D is being made potentially available free for 300000 children in order to deal with 350 presumed cases of rickets within an easily identified ethnic group in Glasgow. Clearly a more specific approach to the vulnerable group, especially teaching English and nutrition to the mothers should be the aim. The problem should be largely self-limited. Today's Asian schoolgirls who know English are tomorrow's mothers and should learn how to avoid this nutritional problem in their offspring. This is a matter for School Health Education.

The long-term solution needs accurate definition of the aetiological factors and of pathogenesis, education of Asians to avoid excessive use of high extract flour and by the use of vitamin D supplements to prevent the development of the illness. Unfortunately health services in Britain have rather frittered away the years since 1962 awaiting effective direction whilst

Asian children have continued to suffer, and are suffering, unnecessarily. A perfect solution may be the aim of all concerned but as Niilo Hallman says 'Pursuit of perfection is the enemy of progress'. All Asian children awaiting osteotomy in Britain in 1979 will echo these sentiments.

CALCIUM DEFICIENCY RICKETS

Dietary inadequacy of calcium has long been recognized as causing rickets in various birds and animals. As early as 1954 Parsons suggested that when the diet of infants was very low in calcium, rickets might occur in babies receiving adequate vitamin D intake. Isolated cases in infants on special diets have been reported in 1970[19] and in 1977[20]. No doubt the purist today might insist on the demonstration of normal levels of serum 25-HCC prevailing before accepting inadequate calcium intake as a cause of rickets but inevitably if such evidence were forthcoming normal levels of serum 1,25-DHCC would then be considered mandatory! Fortunately such evidence is now coming to hand.

Several factors contrive to complicate the picture. Firstly in the past it has been common for children with recognized rickets to receive treatment not only with vitamin D supplements but also with calcium supplements given either deliberately or incidentally as milk forming part of the hospital diet or general dietetic supervision of the affected infant. Furthermore, calcium deficiency rickets in animals tends to be found in species with rapid growth rates compared to man, that is in lambs, pigs and calves which all reach puberty in ½-2 years as opposed to 12-14 years for the human. In milk-drinking countries, whether underdeveloped, developed or overdeveloped, the required intake of calcium daily is easily met. In countries where little or no milk is consumed the intake may be very low. Calcium requirements are still not accurately defined but results from some developing countries suggest that on as little as 200-300 mg of calcium daily no ill effects are obvious. It seems highly likely that other factors may be relevant such as concurrent phosphorus intake, ingestion of vitamin D and availability of sunshine, and skyshine and degree of skin exposure.

Taitz in 1962[21] at Baragwanath Hospital in South Africa identified a group of older black children with rickets and with no underlying disorders such as hyperphosphaturia. These children were usually rurally domiciled and were noted to have mild hypocalcaemia, normal serum phosphorus levels and to heal spontaneously on 'normal' hospital diet.

In 1978 Pettifor and co-workers[22] reported nine black children with such rickets seven of whom were girls. The calcium intake in such black Africans was reckoned to range from 174-475 mg/day. In hospital a diet containing at least 944 mg calcium daily was given without extra vitamin D and all children healed on this regime. None of the children had overt malnutrition although all were below the third centile for height. Legs were mainly affected with knock-knees (genu valgum) or windsweeping. Serum calcium levels ranged from 1.6-2.4 mmol/l, phosphorus from 3.7-5.8 mg/100 ml, alkaline phosphatase from 412–1730 IU/l and 25-HCC from 15–24.5 ng/ml. Secondary hyperparathyroidism was indicated by high urinary cyclic AMP excretion and reversible generalized amino-aciduria. No other illness liable to cause rickets was detected in these children and there was no evidence of either lack or excess of fluoride or of excess of strontium in the diet to inhibit 1-α-hydroxylation of 25-HCC. There was no evidence of any blocking mechanism for calcium absorption and the biochemical abnormalities reversed when calcium was added to the diet with healing of the lesions. They concluded that lack of calcium intake rather than an inhibitory mechanism was the most likely mechanism.

Since then Pettifor et al.[23][24] have pursued the subject and I am deeply grateful for personal communications which they have allowed me to quote. In the first study the prevalence of biochemical abnormalities usually associated with rickets was measured in three black school populations from rural, small town and city environments. No hypocalcaemia was found in city dwellers but 13.2% of rural children were hypocalcaemic and 41.5% had elevated serum alkaline phosphatase concentrations. Hypocalciuria was also present in those from the rural community. Dietary calcium intake in those children with biochemical abnormalities was estimated as 125 mg daily in contrast to an average of 337 mg daily in children with normal biochemistry. These findings strongly support the view that inadequate calcium intake is causal especially since all children had serum 25-HCC concentrations well above the deficient range of 10 ng/ml. This work will shortly be published in full.

The second study related to 13 black children aged 4–15 years with biochemical and radiological rickets. On admission to hospital hypocalcaemia and high serum concentration of alkaline phosphatase were present. Mean 25-HCC level was 22.1 ± 6.1 ng/ml and mean serum 1,25-DHCC for five children was 88 ± 15 pg/ml.

Dietary histories of five rachitic children and six controls from the same community indicated a low daily calcium intake (125 ± 40 mg daily) in rachitic subjects and 337 ± 101 mg in controls. On a diet containing 1200 mg calcium in the ward with no vitamin D supplementation the radiological and biochemical findings improved within three weeks and on dismissal (in approximately 12 weeks) the average serum calcium concentration was normal, the alkaline phosphatase concentration had fallen and the serum level of 1,25-DHCC was 48 ± 14 pg/ml.

These findings strongly suggest inadequate daily intake of calcium as the cause of such rickets. This view is confirmed by the results of calcium supplementation (500 mg daily) given to 30 children in their rural environment over a three month period whilst 30 controls received placebo. In the calcium supplemented group the mean serum calcium concentration rose and serum alkaline phosphatase level fell. No such changes occurred in the placebo group.

CONCLUSIONS

What must have become obvious from the above discussion is that the very term rickets is becoming obsolescent. This was obvious from 1968 when we first recognized subclinical rickets (or hypovitaminosis D?) and it has become more obvious still with the progressive merging of neonatal tetany, dental damage and congenital rickets. As more sophisticated measurements of serum concentrations of 25-HCC and 1,25-DHCC make detection of even more subclinical rickets (or hypovitaminosis D?) possible its relevance to prenatal and postnatal growth becomes progressively more relevant. Furthermore if inadequate calcium intake per se can cause rickets may not a combination of lesser degrees of calcium and vitamin D lack be synergistic in causing rickets?

Acknowledgements

My thanks are due to a range of colleagues who have allowed me to use data much of which is unpublished. These include Professor F. Cockburn, Dr M. G. Dunnigan, Dr A. Ford, Dr K. Goel and Dr R. Gardee from Glasgow, Professor and Dr Teotia from Meerut, India and Dr J. M. Pettifor from Johannesburg, South Africa and their colleagues.

References

1. Arneil, G. C. (1977). Rickets in Britain. In: *Child Nutrition and its Relation to Mental and Physical Development*, p. 35. (London: Kellog Company of Great Britain Limited)
2. Maxwell, J. P. and Turnbull, H. M. (1930). Two cases of fœtal rickets. *J. Pathol. Bacteriol.*, 33, 327
3. Ford, J. A., Davidson, D. C., McIntosh, W. B., Fyfe, W. M. and Dunnigan, M. G. (1973). Neonatal rickets in an Asian population. *Br. Med. J.*, 3, 211
4. Moncrieff, M. W. and Fadahunsi, E. D. (1974). Congenital rickets due to maternal vitamin D deficiency. *Arch. Dis. Child.*, 49, 810
5. Teotia, M., Teotia, S. P. S. and Singh, R. K. (1978). Maternal hypovitaminosis and congenital rickets. *Bull. Int. Paediatr. Assoc.* (In press)
6. Indian Council of Medical Research. (1968). Recommended daily allowances of nutrients and balanced diets.
7. Cockburn, F., Barrie, W. J. M., Belton, N. R., Giles, M., Stephen, R., Pocock, S., Purvis, R. J., Brown, J. K., Kirkwood, J. G., Smith, J., Turner, T. L., Wilkinson, E. and Forfar, J. O. (1979). Maternal vitamin D intake and calcium metabolism in mother and newborn infant. (In preparation)
8. Fleischman, A. R., Rosen, J. F. and Nathenson, G. (1978). 25-HCC for early neonatal hypocalcaemia. *Am. J. Dis. Child.*, 132, 973
9. Arneil, G. C. and Crosbie, J. C. (1963). Infantile rickets returns to Glasgow. *Lancet*, ii, 423
10. Richards, I. D. G., Sweet, E. M. and Arneil, G. G. (1968). Infantile rickets persists in Glasgow. *Lancet*, i, 803
11. Goel, K. M., Sweet, E. M., Logan, R. W., Warren, J. M., Arneil, G. C. and Shanks, R. A. (1976). Florid and sub-clinical rickets among immigrant children in Glasgow. *Lancet*, i, 1141
12. Goel, K. M., House, F. and Shanks, R. A. (1978). Infant-feeding practices among immigrants in Glasgow. *Br. Med. J.*, 3, 1181
13. Dunnigan, M. G., Paton, J. P. J., Haase, G., McNicol, G. W., Gardner, M. D. and Smith, C. K. (1962). Late rickets and osteomalacia in the Pakistani community in Glasgow. *Scot. Med. J.*, 7, 159

14. Dunnigan, M. G. (1977). Asian rickets and osteomalacia in Britain. In: *Child Nutrition and its Relation to Mental and Physical Development*, p. 43. (London: Kellogg Company of Great Britain Limited)
15. Pietrek, J., Windo, J., Preece, M. A., O'Riordan, J. L. H., Dunnigan, M. G., McIntosh, W. B. and Ford, J. A. (1976). Prevention of vitamin D deficiency in Asians. *Lancet*, **i**, 1145
16. Goel, K. M. (1979). A nutrition survey of immigrant children in Glasgow (1974–1976). *Scottish Health Service Studies*, **No. 40**. (Edinburgh: Scottish Home and Health Department) (In press)
17. Lakdawala, D. R. and Widdowson, E. M. (1977). Vitamin D in human milk. *Lancet*, **i**, 167
18. Gardee, R. (1978). An appraisal of the problems of 'Asian' rickets and osteomalacia in Glasgow. (Paper presented to Greater Glasgow Area Health Board)
19. Maltz, H. E., Fish, M. B. and Holliday, M. A. (1970). Calcium deficiency rickets and the renal response to calcium infusion. *Pediatrics, (Springfield)*, **46**, 865
20. Kooh, S. W., Fraser, D., Reilly, B. J. (1977). Rickets due to calcium deficiency. *N. Engl. J. Med.*, **297**, 1264
21. Taitz, L. S. (1962). The role of parathyroid glands in vitamin D deficiency rickets. DM Thesis: University of Witwatersrand, p. 132
22. Pettifor, J. M., Wang, J. and Couper-Smith, J. (1978). Rickets in children of rural origin in South Africa: Is low dietary calcium a factor? *J. Pediatr.*, **92**, 320
23. Pettifor, J. M., Ross, F. P., Moodley, G., DeLuca, H. F., Travers, R. and Glorieux, F. H. (1979). Calcium deficiency rickets. (In press)
24. Pettifor, J. M., Ross, F. P., Moodley, G. and Sheunyane, E. (1979). Calcium deficiency in rural black children in South Africa — a comparison between rural and urban communities. (In preparation)

17
Food enrichment

A. E. BENDER

The addition of nutrients to foods can be considered either as a public health problem, where the enrichment is compulsory, or a voluntary enrichment of proprietary foods by the manufacturer.

The latter may or may not be necessary: that is, it may fulfil a real need, or it may be an advertising gimmick, or it may fall between the two and provide an extra supply of nutrients which at worst is harmless and at best serves as an insurance policy.

So far as infant and toddler foods are concerned it would seem that manufacturers are filling a need since it has been noted that if this were not already being done on a voluntary basis there would probably have been a need for legislation. There is a recognized need to supplement an infant's diet with vitamins A and D, since cod liver oil was an (unwelcome) welfare food, and with vitamin C, since orange juice was also provided as a welfare food. The uptake of these welfare foods was never universal and the addition of vitamins to infant foods has probably made these vitamins more widely available.

So far as ordinary adult foods are concerned, nutritionists tend to fall into two camps, those who believe that extra vitamin supplements are not necessary since the diet can provide all that is needed, and those who suggest that supplements serve as an insurance policy since we do not know whether the full needs of all the population are being met. When the manufacturer enriches his products he is supplying the extra, quite apart from the rather separate question of replacing processing losses.

PUBLIC HEALTH ENRICHMENT
So far as public health measures are concerned the procedure should follow a

163

logical progress (although historically, this has not always occurred): (1) demonstrate the need to enrich; (2) establish the best vehicle to carry the nutrients; (3) ensure that there is no detriment to palatability and acceptability and that the cost is acceptable; (4) make available the required technology; and (5) have the machinery for legal enforcement.

Enrichment is often desirable in developing countries but cannot be carried out because one or more of these stages presents an obstacle.

(1) Demonstration of need would probably be essential at the present time in any country despite what may have happened in the past since there is considerable cost involved.

Enrichment of white bread in the United States followed reports of inadequate *intake* of thiamin in a large part of the population but there were no signs of deficiency. The addition of vitamin A to margarine in the UK was rather a moral point since the rationing of butter during the Second World War effectively deprived the consumers of part of their vitamin intake which might have had nutritional repercussions. The outbreak of xerophthalmia in Denmark in 1916, consequent upon the export of butter and its replacement with unenriched margarine was an obvious warning.

(2) The vehicle must be a food commonly consumed, especially by the population groups in greatest need.

(3) Enrichment cannot be carried out unless the nutrient is compatible with the food, although the preparation of water-dispersible forms of fat-soluble vitamins has overcome part of this problem. There must obviously be no change in colour, flavour or texture. This militates for example, against high level fortification with iron since this can form coloured complexes as well as promote oxidation. Since vitamins are required in such small quantities related to the foods themselves they generally have no detectable effect on appearance or flavour whereas protein enrichment presents obvious problems.

(4) The technology has posed considerable problems in the past, such as how to blend very small amounts of vitamin mixtures with large volumes of materials of different density, or how to add amino acids, vitamins or minerals to whole cereal grains. Even when the problems have been solved the application often presents a major obstacle to enrichment in developing countries. One of the greatest obstacles often lies in the very large number of small-scale processing plants where it would not be possible to control the additions; when families rely largely or solely on foods that they grow themselves then enrichment cannot be carried out.

(5) Finally, it is little use passing regulations if they cannot be enforced. Enrichment by law requires the machinery for sampling foods, for analysing large numbers of samples in many different areas of the country, and, finally, of enforcing a penalty in the courts. Much potential enrichment has foundered on this point in countries where the population would clearly have benefited from enrichment.

Restoration

Restoration is the most widespread of the enrichment programmes as exemplified by white bread in many countries (Table 1); very few countries compel enrichment of rice. Even countries that do not permit general addition of nutrients to foods do allow restoration. As the table shows there is some variation in the number and amount of nutrients added because of national policies. There are, in addition, proprietary cereal products with varying amounts of various nutrients.

Table 1 Enrichment of cereal products (white flour except where otherwise stated) (per kg)

	Thiamin	Riboflavin	Nicotinic acid	Iron	Calcium
	(mg)	(mg)	(mg)	(mg)	(mg)
Australia	1.6	2.4	16	14.7	1000
*Brazil	4.5	2.5	—	30	1000
Canada	4.4–5.5	2.7–3.3	35–44	29–36	1100–1400
*Chile	6.3	1.3	13	13.3	1700
(rice)	4.4–8.8	2.6–5.3	35–70	29–57	1100–1650
Congo (Dem. Rep.)	4–6	2.5–3.5	32–45	26–35	1000–1500
Costa Rica	4.4–5.5	2.6–3.3	35–44	29–44	1100–1400
*Denmark	5	5	—	30	5000
*(rye flour)	—	—	—	30	10000
Dominica	4.4–5.5	2.6–3.3	35–44	29–36	1100–1400
Germany	3–4	1.5–5.0	20	30	720–2000
Israel	—	2.5	—	—	—
Japan	5	3	—	—	1500
Nicaragua	1	1.4	15.7	13	500
*Panama	4.4	2.6	35	28.7	1100
*Peru	4.0	4.0	30	20	1000
*Philippines	4.4–5.5	2.6–3.3	35–44	29–36	1100–1400
Portugal	4.4–5.5	2.6–3.3	35–44	28–36	—
*Puerto Rico	4.2	2.4–2.5	30	26–36	1100
Sweden	2.6–4.0	1.2	23–40	30	—
Switzerland	2.8–4.2	1.7–2.5	29–44	18–26	—
*United Kingdom	2.4	—	16	16.5	1250
†United States					
white flour	4.4–5.5	2.6–3.3	35–44	29–36	1100–1400
bread	2.4–4.0	1.6–3.5	22–33	18–28	660–1750
corn meal	4.4–6.6	2.6–4.0	35–53	29–57	1100–1600
rice	4.4–8.8	2.6–5.3	35–70	29–57	1100–2200
pastas	8.8–11.0	3.7–4.8	60–75	29–36	1100–1400
USSR	2–4	4	10–30	—	—

* Legally enforced. (*Note:* Some of the information in this table is a compromise conflicting reports.)
† Legal enforcement in 30 States (vitamin D also added 8–50 µg/kg)

In Great Britain and Canada the nutrients are added to the flour, so that all bakery products are enriched. In Great Britain the law demands that white flour, mostly 70% extraction rate, should be enriched at the rate of 14 oz of creta preparata (treated calcium carbonate) to the standard 280 lb sack, which is equivalent to about 90 mg calcium/ 100 g bread, and not less than 1.65 mg iron, 0.24 mg thiamin and 1.6 mg nicotinic acid/ 100 g flour. All flour other than wholemeal (100% extraction) must be enriched.

In the United States enriched flour must contain not less than 0.44 mg thiamin, 0.26 mg riboflavin, 3.6 mg nicotinic acid and 2.9 mg iron/ 100 g of the flour; calcium is not specified. Although the levels set for bread are related to the flour and the 1943 Order of the US War Food Administration applied, at that time, to the bread, in practice it is the dough that is enriched by the baker.

The other food that is commonly enriched for public health reasons is margarine (see Table 2). It will be observed that enrichment is mainly carried out in the developed countries.

Table 2   Enrichment of margarine (per kg)

	Vitamin A (μg)	Vitamin D (μg)
Australia	9000	100
Austria	6000	25
Belgium	6000	25
Brazil	4500–15000	12.5–50
Canada	10000	—
Chile	9000	25
Denmark	6000	15
Finland	6000	62–90
Germany	6000–9000	7–25
Greece	7500	37
India	7500	—
Israel	9000	75
Japan	6000–12000	—
Mexico	6000	50
Netherlands	6000	25
Norway	6000	62
Portugal	6000–10000	22–25
Sweden	9000	37
Switzerland	9000	75
South Africa	6000	25
Turkey	6000	25
United Kingdom	9000	70–90
United States	10000	110

The history of enrichment of flour in Great Britain illustrates some of the reasons why the logical stages may not be followed. In 1941 synthetic thiamin became available on a large scale and it was decided to add this to the white flour of 70–72% extraction rate. Before this policy could be implemented large shipping losses of wheat reduced stocks to a dangerously low level and it was decided to raise the extraction rate to 85% as an economy measure (and at times to 95%). This made fortification unnecessary but introduced the problem that the higher phytate content might reduce the availability of calcium especially as milk was in limited supply: (zinc was not known to present a problem at that time). Consequently calcium was added in the form of calcium carbonate at the rate, first of 7 oz, then of 14 oz per 280 lb sack of flour. With so high an extraction rate no further enrichment was necessary.

In 1952 when the controls on the milling of wheat were lifted there was said to be a popular desire to return to white flour and it was decided to enrich this with 'token' nutrients up to the level of the 80% extraction flour since this had apparently contributed towards general good health in the wartime years. So the post-war loaf, other than wholemeal, was enriched with iron, thiamin, nicotinic acid and calcium. (The 1974 Bread and Flour Report revealed new values for the vitamin content of 80% extraction flour.)

A recommendation of the Food Standards Committee in 1974 stated that there was no problem of niacin deficiency among the British population since this is made in the body from tryptophan, and so nicotinic acid could safely be omitted from the enrichment programme. Calcium was also considered to be unnecessary since there were adequate amounts in the diet but the Committee was impressed with the observation that coronary heart disease was less common in areas where the water supply was hard, i:e. richer in calcium, than in soft water areas, and until this has been explained the Committee thought it safer to continue enriching with calcium.

Voluntary enrichment

A wide variety of manufactured foods are voluntarily enriched. For example most, if not all, infant milk formulae contain a range of vitamins and minerals and most infant cereal products contain added thiamin, riboflavin and nicotinic acid, vitamin D, calcium and iron. Some include vitamin A and others have a complete range including B_6, B_{12}, folate and pantothenate. The regulations do not permit claims to be made for vitamins other than A, thiamin, riboflavin, niacin, C and D and for minerals other than calcium, iron and iodine.

Many breakfast cereals are enriched, some with a limited range of the vitamins and iron, others with a more complete range, sometimes including vitamin C and additional protein.

One of the earliest nutrients to be added to foods was iodine to prevent goitre. It was suggested as long ago as 1833 by Boussingault and adopted in Switzerland shortly after 1900, when it was added to chocolate. A commoner vehicle for iodine is salt (sodium chloride) but in most countries this is done only on a voluntary basis so that it is available in areas where it is needed rather than compulsorily provided for everyone since problems can arise from indiscriminate fortification with iodine.

Enrichment of milk with vitamin D was started in the United States in 1931 because of the common occurrence of rickets. There are enrichment programmes covering a variety of staple foods in many countries but they are not legally enforced and so, although fostered by governments, they are voluntary. Enrichment programmes for rice have been developed in Colombia, Taiwan, Hawaii, Japan, United States and Venezuela. Maize is enriched in several states of the United States, Egypt, Mexico and Yugoslavia. Margarine enrichment is mandatory in some countires, voluntary in others.

The amino acids lysine and methionine, one or other of which is limiting in almost every food and diet, are available on a large scale and are occasionally used to raise the biological value of the protein of a food.

Some proprietary foods are enriched with protein although it could be argued that any country sufficiently technically developed to do this has no need for extra protein.

DEVELOPING COUNTRIES

There is a clear and urgent need for enrichment in many developing countries, although, because of the practical difficulties, such programmes are rare.

In India, vitamin A deficiency is widespread and one of the remedies is enrichment of tea. Small leaves, tea dust, are enriched by mixing with powdered retinol palmitate and only 15% is lost after 1 year's storage. The stability is probably due to the antioxidants naturally present.

The concentration is 37.5 μg vitamin A/g tea, and since 3 g is used per cup, this should provide approx. 100 μg/cup. An adult drinking three cups/day would thus receive 300 μg towards the recommended daily intake (RDI) of 750 μg, and a child drinking 1–2 cups should receive 100–200 μg.

In Guatemala vitamin A is being added to table sugar; some authorities have considered adding B vitamins to sugar on the principle that it makes a demand on the rest of the diet for the B vitamins needed for its metabolism.

An example of some of the principles and problems involved is provided by the enrichment of atta, the wheat flour used extensively in India for making chappatis, inaugurated in 1970. Most of the 20 million tons of wheat consumed per year is ground in small, hand operated mills, and only three million tons pass through large mills and so can be enriched, as follows (per ton):-

edible grade groundnut flour (45–50% protein), 50 kg,
retinol 9.2 g, riboflavin 1.38 g, nicotinic acid 7.6 g,
thiamin 1.5 g, calcium diphosphate 800 g,
iron as ferrous sulphate 96 g, calcium carbonate 800 g

In 1970 enrichment in this way added 4% to the cost.

PHILOSOPHY OF ENRICHMENT

Some authorities accept enrichment only as a short-term policy pending improved food supplies, improved processing methods resulting in higher retention of nutrients, and nutritional education of the consumer.

Others regard it as a permanent policy and indeed, the Council for Foods and Nutrition of the American Medical Association has stated that efforts to improve the nutritional status of populations by the introduction of new foods and by attempts to change food habits often have not been effective.

Another aspect is the contention that foods should be enriched only with such nutrients as they already contain, although in inadequate amounts. In the United States the view has been expressed that 'to avoid undue artificiality of foods that are enriched, one should select only those that have suffered losses in processing and use only those in type or amount that are normally found in those foods.' On these grounds permission has often been refused to US manufacturers wishing to enrich proprietary foods.

Such views assume that unprocessed foods contain the nutrients required by man in the required proportions, as does the view that the nutritional profile of an unprocessed food must be preserved. It leads to the enrichment of fruit juices with vitamin C despite the fact that consumers of fruit juices will already be obtaining considerable amounts of this vitamin, while it is those who do not take such foods who may well be in short supply. It would therefore be logical to upset the nutritional profile and render some foods 'artificial' by adding the vitamin C to drinks like tea and coffee — such a policy, in fact, as has been accepted by the enrichment of tea with vitamin A in India and the enrichment of sugar.

FUTURE DEVELOPMENTS

Four changes are taking place that will influence future policy in this field — increasing nutritional knowledge, more information about the precise nutrient intake and its range within population groups, a move from traditional towards newer foods and a breakdown in the family eating pattern with greater dependence of some sections of the community on caterers and snack foods.

Guidance of future trends can be sought in American action. The Council on Foods and Nutrition of the American Medical Association classifies foods as conventional, formulated (mixtures of two or more foodstuffs or ingredients processed and blended together including breakfast cereals, convenience foods and snack foods) and fabricated (prepared principally from ingredients specifically designed to achieve a particular function not possible with common food ingredients).

The nutritional value of conventional foods can, they say, be improved by good processing practices. Restoration is acceptable when the product in question is nutritionally important, i.e. the nutrients originally present provided at least 5% RDI in a serving. 'Industry should strive to improve techniques rather than depend on restoration.'

When a formulated food has no easily identifiable counterpart among conventional foods and contributes 5% or more of the RDI of energy or any essential nutrient in one serving then nutrient additions should be related to the energy supplied, i.e. if it supplied 10% of the energy then it should supply the same percentage of the nutrients. If it already contains three-quarters of the nutrient then no adjustment is needed. A product designed primarily as a meal replacer should provide 25–50% of all nutrients listed in RDI tables (except energy). Prepared breakfast cereals may be designed to provide up to 25% RDI per ounce, even if they are low energy foods for weight loss.

Fabricated foods may or may not closely resemble existing foods. If they do they should contain on an energy basis at least the variety and quantity (at least 75%) of nutrients contained in important amounts in the general class of foods to which the imitated food belongs.

A new food should provide one to one and a half times the representative RDI for important nutrients in relation to the energy supplied. 'Selection of nutrients and the amounts used should be based upon the producer's suggested pattern of use of the food by consumers, the stability of the nutrients in the foods and good manufacturing practice.' (Council on Foods and Nutrition, American Medical Association.)

Exceptions to these guidelines are made for dietetic foods. The Report of the British Food Standard Committee on Novel Protein Foods in 1974 points in the same direction. It is clearly implied here that any new foods which are expected to replace traditional foods shall supply at least as much of the nutrient as the foods they replace. The obvious example is textured vegetable protein food intended as an alternative to meat.

Since meat supplies a valuable proportion of the thiamin, niacin and vitamin B_{12}, and since the iron of meat is particularly well absorbed in contrast with the poor absorption from most other foods, such novel foods, it is suggested, should be enriched with these nutrients. The principle has long been applied to margarine.

Similar considerations will doubtless apply to all new products. Instant potato replaces a valuable source of vitamin C in many diets in the western countries and most manufacturers restore or even 'nutrify' with vitamin C.

It is more likely that the solution to these potential problems lies with the voluntary enrichment by manufacturers of proprietary foods than with legislation. This is partially because the problem may be insufficeintly clear to require official action and partially because it is not possible to legislate for new foods that are not intended to replace any particular traditional food. While legislation could easily be effected to ensure that meat replacers supplied most of the nutrients previously supplied by meat it is difficult to visualize what, if any, standards could be set for new products that do not claim to replace any particular article of diet. Fortunately market competition will almost certainly fill any nutritional requirements, apart from the manufacturer's sense of responsibility towards his customers' nutritional needs.

Enrichment even of individual manufactured foods can have a marked effect on large sections of the population. Marr and Berry[1] drew attention to the effect of enrichment of a manufactured food on the average national intake of riboflavin. They stated that riboflavin intake was rising slowly over the years up to 1965, with increasing consumption of animal foods. Then there was a sharp rise in 1966 which affected all social groups 'due to the decision by a major producer of breakfast cereals to fortify products with riboflavin'. The question of food enrichment has been reviewed in greater detail by Bender[2].

References

1. Marr, J. W. & Berry, W. T. C. (1974). *Nutrition*, **28**, 39
2. Bender, A. E. (1978). *Food Processing and Nutrition.* (London: Academic Press)

F. L. CALDER CAMPUS

Index

173